Pediatric Epilepsy Case Studies

From Infancy and Childhood through Adolescence

Pediatric Epilepsy Case Studies

From Infancy and Childhood through Adolescence

Edited by

Kevin Chapman, MD
Barrow Neurological Institute, Phoenix, AZ

Jong M. Rho, MD
Barrow Neurological Institute, Phoenix, AZ

CRC Press
Taylor & Francis Group
Boca Raton London New York

CRC Press is an imprint of the
Taylor & Francis Group, an **informa** business

Cover illustration courtesy of the Barrow Neurological Institute (© 2007, Barrow)

CRC Press
Taylor & Francis Group
6000 Broken Sound Parkway NW, Suite 300
Boca Raton, FL 33487-2742

© 2009 by Taylor & Francis Group, LLC
CRC Press is an imprint of Taylor & Francis Group, an Informa business

No claim to original U.S. Government works
Printed in the United States of America on acid-free paper
10 9 8 7 6 5 4 3 2 1

International Standard Book Number-13: 978-1-4200-8342-2 (Hardcover)

Library of Congress Cataloging-in-Publication Data

Pediatric epilepsy case studies : from infancy and childhood through adolescence / editors, Kevin Chapman, Jong M. Rho.
    p. ; cm.
"A CRC title."
Includes bibliographical references and index.
ISBN 978-1-4200-8342-2 (hardback : alk. paper)
1. Epilepsy in children--Case studies. I. Chapman, Kevin, MD. II. Rho, Jong M.
[DNLM: 1. Epilepsy--Case Reports. 2. Adolescent. 3. Child. 4. Infant. WL 385 P377 2008]

RJ496.E6P42 2008
618.92'853--dc22
                                        2008022878

Visit the Taylor & Francis Web site at
http://www.taylorandfrancis.com

and the CRC Press Web site at
http://www.crcpress.com

Contents

SECTION 1: THE BASICS

SECTION 2: THE NEONATE

SECTION 5: THE ADOLESCENT

Dedication

To our patients and their families,
who continue to teach and challenge us
to improve our treatment of epilepsy

Foreword

Whenever a child presents with a seizure, physician, family and patient wonder about recurrence and whether it is epilepsy. The seizure type and possible epilepsy syndrome is to be delineated by careful clinical assessment and probable ancillary testing. The exact clinical approach to each patient depends upon the situation in which a seizure or seizures occurred, the associated factors, description of the events, and the child's comorbid conditions. Whether an acute illness, underlying neurologic or mental handicap, or seemingly progressive deterioration was present will determine whether to evaluate the child emergently or in the more routine fashion. Treatment decisions follow in the hope of stopping all seizure recurrence, producing no deleterious effects, and allowing as normal development as possible.

This book is intended to give practical information regarding the diagnosis and management of children with epilepsy through a case-based approach. Following a section entitled "The Basics," the editors have assigned child neurology experts to discuss various seizure, epilepsy, and disease entities so that the readers can adequately evaluate and form a treatment plan for each patient type. An age-based approach allows the reader to consider the appropriate conditions possibly presenting in their patient, and each clinical vignette discusses the essential characteristics, incorporating the differential diagnosis that should be considered.

This case-based approach is, in fact, the way most clinicians best learn to differentiate and manage various medical conditions. Epilepsy is no different. Indeed, we are all "students," each day learning about the nuances of pediatric epilepsy, its similarities and differences. Newer techniques of imaging, biochemical and genetic testing, and potential therapies through medication, surgery, and diet are all evolving. Each of us will find these case descriptions and discussions informative while reminding the reader of a "clinical pearl" or remembering "that case I saw... ." Seasoned clinicians will appreciate important lessons while reading about new techniques, while the novice medical professional will incorporate both basics and advanced understanding of pediatric epilepsy to use regarding their current and future patients. These case examples allow the clinician to appreciate the importance of establishing the epilepsy syndrome.

Although many will not read this text from cover to cover, my conviction is that most will refer to it many times, as they review the clinical scenario which best fits their individual patient or while they look for a specific test or find a reference regarding an entity that they suspect in their patient. All of us have and will continue to learn through case presentation and example. The editors and authors have indeed provided a real service in *Pediatric Epilepsy Case Studies: From Infancy and Childhood through Adolescence*.

John M. Pellock, M.D.
Professor and Chairman, Division of Child Neurology
Virginia Commonwealth University/Medical College of Virginia Hospitals
Richmond

Preface

Epilepsy encompasses a wide variety of clinical syndromes characterized by heterogeneous etiologies, presentations, and prognoses. Accurate diagnosis is critically important for the proper care of patients with epilepsy, especially since some forms of epilepsy have a benign course whereas others are associated with progressive neurocognitive decline. Advances in neuroimaging and genetics have improved our diagnostic abilities and our fundamental understanding of the epilepsies. In addition, newer medications have offered patients better tolerability than traditional agents, but unfortunately, no significant improvements in overall seizure control have been afforded. Many epileptic conditions remain intractable to currently available medications. However, other nonpharmacological treatment options (such as the ketogenic diet and vagus nerve stimulator) may provide some hope for improved seizure control in these patients with medically refractory epilepsy.

For the physician in training, grasping the complexity and nuances associated with various epileptic syndromes can be daunting. The goal of this book is to help students, residents, and healthcare professionals understand the different epilepsies encountered in clinical practice across the pediatric age range. The initial section provides an introduction to the fundamentals of epilepsy, and subsequent sections include succinct case presentations and clinically relevant discussions of the more common epilepsy syndromes affecting each age group. Suggested references are also provided to guide the reader toward more detailed studies of a specific topic of interest.

This book is the culmination of a group effort involving many of the leading physicians and researchers in the field of pediatric epilepsy. We believe that their individual contributions together constitute a concise and practical reference for health professionals in training. Research in the field of epilepsy continues at a rapid pace, with the ultimate hope of curing many intractable epilepsy patients. We hope that this book may spark the interest of residents, trainees, and other healthcare professionals in joining the international fight against epilepsy.

Kevin Chapman, M.D.
Barrow Neurological Institute

Jong M. Rho, M.D.
Barrow Neurological Institute

Editors

Dr. Kevin Chapman completed his residency training at Baylor College of Medicine and a clinical neurophysiology fellowship at the Cleveland Clinic Foundation. He served as faculty at Baylor College of Medicine before becoming faculty at the Barrow Neurologic Institute. Dr. Chapman's clinical and research areas of interest involve the surgical management of epilepsy, medical treatment of hypothalamic hamartoma, and the ketogenic diet.

Dr. Jong M. Rho completed training in pediatric neurology at UCLA Medical Center and in neuropharmacology at the National Institutes of Health. He served as a faculty member at the University of Washington (Seattle Children's Hospital and Regional Medical Center) and the University of California at Irvine before assuming his position at the Barrow Neurological Institute, where he is the associate director of child neurology. Dr. Rho's research interests involve basic mechanisms of ketogenic diet action, developmental animal models of epilepsy, and laboratory studies of surgically-resected human epileptic tissue.

Kevin Chapman, M.D.
Division of Pediatric Neurology
St. Joseph's Hospital and Medical Center
Barrow Neurological Institute
Assistant Professor of Clinical Pediatrics and Neurology
University of Arizona
College of Medicine-Phoenix
Phoenix, Arizona

Harry T. Chugani, M.D.
Carman and Ann Adams Department of Pediatrics
Department of Neurology
Children's Hospital of Michigan
Wayne State University
Detroit, Michigan

Dave F. Clarke, M.D.
Department of Pediatrics
Division of Pediatric Neurology
University of Tennessee Health Science Center
Le Bonheur Comprehensive Epilepsy Program
Memphis, Tennessee

Edward C. Cooper, M.D., Ph.D.
Penn Epilepsy Center
Hospital of the University of Pennsylvania
Department of Neurology
University of Pennsylvania
Philadelphia, Pennsylvania

Cornelia Drees, M.D.
Department of Neurology
Barrow Neurological Institute
St. Joseph's Hospital and Medical Center
Phoenix, Arizona

L. Matthew Frank, M.D.
Associate Professor of Neurology and Pediatrics
Eastern Virginia Medical School
Departments of Pediatrics and Neurology
Children's Hospital of The King's Daughters
Norfolk, Virginia

James D. Frost, Jr., M.D.
Professor of Neurology and Neuroscience
Baylor College of Medicine
Houston, Texas

Christopher C. Giza, M.D.
UCLA Brain Injury Research Center
Division of Pediatric Neurology and Department of Neurosurgery
Mattel Children's Hospital at UCLA
David Geffen School of Medicine at UCLA
Los Angeles, California

Ajay Gupta, M.D.
Pediatric Neurology and Epilepsy Center
Department of Neurology, Neurological Institute
Cleveland Clinic Foundation
Cleveland, Ohio

Richard A. Hrachovy, M.D.
Professor of Neurology
Baylor College of Medicine
The Michael E. DeBakey Veterans Affairs Medical Center
Houston, Texas

John F. Kerrigan, M.D.
Assistant Professor of Clinical Pediatrics and Neurology
University of Arizona College of Medicine-Phoenix
Director, Pediatric Epilepsy Program
Barrow Neurological Institute and Children's Health Center
St. Joseph's Hospital and Medical Center
Phoenix, Arizona

Susan Koh, M.D.
Associate Professor of The Children's Hospital of Denver
University of Colorado
Denver, Colorado

Eric H. Kossoff, M.D.
Assistant Professor of Pediatrics and Neurology
Medical Director, Ketogenic Diet Program
John M. Freeman Pediatric Epilepsy Center
Johns Hopkins Hospital
Baltimore, Maryland

Michael C. Kruer, M.D.
Department of Pediatrics
Divisions of Pediatric Neurology and Developmental Pediatrics
Oregon Health and Science University
Portland, Oregon

Paul M. Levisohn, M.D.
Associate Professor of Pediatrics and Neurology
University of Colorado at Denver
Health Sciences Center
The Children's Hospital
Aurora, Colorado

Tobias Loddenkemper, M.D.
Division of Epilepsy and Clinical Neurophysiology
Children's Hospital
Boston, Massachusetts

Aimee F. Luat, M.D.
Carman and Ann Adams Department of Pediatrics
Department of Neurology
Children's Hospital of Michigan
Wayne State University
Detroit, Michigan

Eric Marsh, M.D., Ph.D.
Division of Child Neurology
Children's Hospital of Philadelphia
Department of Neurology
University of Pennsylvania
Philadelphia, Pennsylvania

Berge A. Minassian, M.D., C.M., FRCP(C)
Associate Professor and Canada Research Chair in Pediatric Neurogenetics
The Hospital for Sick Children
The University of Toronto
Toronto, Canada

Yu-tze Ng, M.D., FRACP
Division of Pediatric Neurology
Barrow Neurological Institute
Phoenix, Arizona

Douglas R. Nordli, Jr., M.D.
Lorna S. and James P. Langdon
  Chair of Pediatric Epilepsy
Children's Memorial Hospital
Children's Memorial Epilepsy Center
Chicago, Illinois

James W. Owens, M.D., Ph.D.
Assistant Professor of Pediatrics and Neurology Department
Texas Children's Hospital
Baylor College of Medicine
Houston, Texas

Kristen L. Park, M.D.
Pediatric Epilepsy Fellow
Children's Memorial Hospital
Children's Memorial Epilepsy Center
Chicago, Illinois

Jong M. Rho, M.D.
St. Joseph's Hospital and Medical Center
Barrow Neurological Institute
Phoenix, Arizona

James J. Riviello, Jr., M.D.
George Peterkin Endowed Chair in Pediatrics
Department of Pediatrics
Section of Neurology and Developmental Neuroscience
Baylor College of Medicine
Houston, Texas

Colin M. Roberts, M.D.
Doernbecher Childhood Epilepsy Program
Departments of Pediatrics and Neurology
Oregon Health and Science University
Portland, Oregon

Russell P. Saneto, D.O., Ph.D.
Division of Pediatric Neurology
Children's Hospital and Regional Medical Center
University of Washington
Seattle, Washington

Raman Sankar, M.D., Ph.D
Professor and Chief of Pediatric Neurology
Rubin Brown Distinguished Chair
Mattel Children's Hospital at UCLA
David Geffen School of Medicine at UCLA
Los Angeles, California

Mark S. Scher, M.D.
Division Chief of Pediatric Neurology
Rainbow Babies and Children's Hospital
University Hospitals
Case Medical Center
Cleveland, Ohio

W. Donald Shields, M.D.
David Geffen School of Medicine at UCLA
University of California at Los Angeles
Los Angeles, California

Asit K. Tripathy, M.D.
Pediatric Neurology and Epilepsy Center
Department of Neurology
Neurological Institute
Cleveland Clinic Foundation
Cleveland, Ohio

Matthew M. Troester, D.O.
Division of Pediatric Neurology
Barrow Neurological Institute
Children's Health Centers
St. Joseph's Hospital and Medical Center
Phoenix, Arizona

James W. Wheless, M.D.
Professor and Chief of Pediatric Neurology
LeBonheur Chair in Pediatric Neurology
University of Tennessee Health Science Center
Director, Neuroscience Institute and LeBonheur Comprehensive Epilepsy Program
  LeBonheur Children's Medical Center
Clinical Chief and Director of Pediatric Neurology
St Jude Children's Research Hospital
Memphis, Tennessee

Angus A. Wilfong, M.D.
Associate Professor of Pediatrics and Neurology
Baylor College of Medicine
Medical Director, Comprehensive Epilepsy Program
Texas Children's Hospital
Houston, Texas

Korwyn Williams, M.D., Ph.D.
Division of Pediatric Neurology
Phoenix Children's Hospital
Pheonix, Arizona

Mary L. Zupanc, M.D.
Professor of the Department of Neurology and Pediatrics
Chief, Division of Pediatric Neurology
Co-Director, Pediatric Neurology Residency Training Program
Director, Pediatric Comprehensive Epilepsy Program
Heidi Marie Bauman Chair of Epilepsy
Medical College of Wisconsin
Children's Hospital of Wisconsin
Milwaukee, Wisconsin

Section 1

The Basics

1 A Pediatric Epilepsy Primer

James W. Owens, M.D., Ph.D.

CONTENTS

Epilepsy is a common, and commonly misunderstood, chronic medical condition of childhood. As frequently encountered as childhood asthma, convulsions occur in approximately 5% of all children in the United States, and 1% of children are diagnosed with epilepsy.[5,9,11,23] Appropriate diagnosis and management are crucial given the potential for lifelong consequences to the developing brain. Not all seizures need to be treated, as there are differences in the management of true epileptic conditions versus reactive or isolated seizures. An example of the former would be recognizing the clinical phenotype of infantile spasms, a particularly devastating type of developmental brain disorder. An illustration of the latter would be refraining from use of anticonvulsant medications for children with recurrent, brief febrile seizures. This common form of acute provoked seizure does not reflect an enduring epileptic condition, and daily preventative treatment is not warranted. Furthermore, there are many paroxysmal disorders affecting children, such as parasomnias and behavioral problems, which are frequently mistaken for epileptic phenomena. The epilepsies of childhood differ significantly both from each other as well as from those encountered in adulthood; the pediatric brain is not just a smaller adult brain. The key, then, is to understand what epilepsy is and what it is not, and to appreciate the unique age- and syndrome-dependent nature of epileptic conditions to guide proper diagnosis and management. In this introductory chapter, a few key points regarding pediatric epilepsy will be highlighted, and expanded upon in the remainder of this book.

PEDIATRIC EPILEPSY IS COMMON

Within the first two decades of life, approximately 5% of children will have experienced some form of convulsion. A significant majority of these seizures will be acute provoked events, often in the context of a febrile illness, and not the recurrent unprovoked seizures that are the hallmark of epilepsy. Among all children who have a single unprovoked seizure, only about 40% of them will ever have a second.[22] This rate of recurrence varies greatly depending on such factors as what type of seizure occurred and whether there is other evidence of neurological dysfunction. For example, a patient who, at baseline, has an abnormal neurological examination, abnormal electroencephalogram, and abnormal MRI may have a risk of recurrence of approximately 90%.[18] Of course, this does not indicate when a subsequent seizure might actually occur.

Approximately 20% of patients experiencing a convulsion of some type will later develop epilepsy: by 20 years of age, approximately 1% of the population will have been diagnosed with this condition.[11] Published studies of incidence vary greatly, which may be partly due to the inclusion of single unprovoked seizures as well as acute symptomatic seizures in some studies. With respect to age-specific incidence, it seems clear that the onset of epilepsy most frequently occurs at the two extremes of the lifespan. A number of studies have shown that the incidence of epilepsy is high in the first year of life, lowest in middle age, and rises again in the elderly. In a population of patients aged 70 years or more, the incidence is as high as 3%.[11] As one might imagine, the causes, types, and outcomes differ significantly between these two populations, although there is certainly some overlap.

PEDIATRIC EPILEPSY ENCOMPASSES A WIDE RANGE OF DISORDERS

Imagine sitting in the waiting room of a pediatric epilepsy clinic and observing the variety of patients awaiting their turn to be evaluated. A 6-year-old child, initially referred for "staring spells," is now here for a follow-up appointment with well-controlled childhood absence epilepsy. In a wheelchair you see a 13-year-old child with spastic quadriparetic cerebral palsy, moderate mental retardation, and poorly controlled symptomatic localization-related epilepsy. She is here to have the settings on her vagus nerve stimulator adjusted in the hope that her secondarily generalized seizures might become less frequent. An 8-month-old infant has been worked into the schedule, with continued clusters of infantile spasms despite completing a trial of adrenocorticotropic hormone (ACTH). A new patient is here for a second opinion about whether or not his brief stereotyped events of generalized shaking with partial loss of consciousness are epileptic in nature. Finally, there is an 8-year-old for a 6-month postoperative follow-up visit after a focal neocortical resection to remove an area of cortical dysplasia, who happily remains seizure free. As different as these patients may be in age, clinical phenomenology, and response to therapy, they all have epilepsy. Clearly, this is a heterogeneous collection of distinct disorders, which may more appropriately be referred to as "the epilepsies." To understand this clinical spectrum, one must be familiar with both what unifies these conditions and what makes each distinct.

The definition of epilepsy is deceptively simple: having two or more unprovoked seizures separated by more than 24 hours. Each component of that definition is important to bear in mind. Seizures are paroxysms of abnormally hyperexcitable and hypersynchronous cortical activity that result in a change in sensation, motor function, behavior, or the sensorium. If the seizure occurs immediately following a precipitating event, then it is referred to as an acutely provoked/reactive seizure or acute symptomatic seizure. As mentioned previously, a common example of such an event would be a febrile seizure: 2 to 4% of all children between the ages of 6 months and 5 years experience a generalized seizure lasting less than 15 minutes in association with a fever not caused by a central nervous system (CNS) infection. In this case, the acute provoking event—the fever—is immediately followed by the seizure. Other examples of acute symptomatic seizures would include those that occur at the time of trauma, in the context of hyponatremia, or in association with a withdrawal syndrome (e.g., alcohol). In contrast, with epilepsy, there is no immediate provoking event for the seizure. At times, the seizure may arise from an old injury such as from a prior stroke. Because the precipitating event precedes the seizure by weeks to years, such an event is considered unprovoked and is often referred to as a "remote symptomatic seizure." Finally, in order to meet the definition of epilepsy, two or more unprovoked seizures must be separated by more than 24 hours. The reason for this is that rapidly recurrent seizures occurring close together carry the same epidemiological risk of eventual recurrence as a single seizure.[12]

If a patient has two or more unprovoked seizures, then that patient may justifiably be labeled as having epilepsy. Given the broad nature of this definition, many different types of clinical phenotypes fall under the cover of this one large umbrella. It is a bit like stating that one lives in North America—helpful information but not very specific! Nowhere is this more evident than in the pediatric epilepsies in which cause, clinical phenomenology, and outcome vary greatly. As with the epilepsies that arise in adulthood, children may suffer seizures as a consequence of trauma, CNS infections, strokes, and other brain insults. A particular example of this would be children who suffer injuries in utero or during the process of birth. Largely unique to childhood are seizures that arise from developmental brain malformations such as disorders of neuronal migration leading to focal cortical dysplasias. It is interesting to note, however, that although the abnormally formed cortex is present from birth, an epileptic disorder may not develop for many years. The reasons for this remain unclear.

To bring some semblance of order to this landscape, the epilepsies have historically been categorized or classified on the basis of electroclinical features. Clinically, this is accomplished using a schema developed by the International League Against Epilepsy (ILAE), which utilizes etiology and seizure type.[4,6] If the patient's epilepsy arises from an evident cause, such as a remote symptomatic seizure due to an old brain injury, then the epilepsy is referred to as *symptomatic* (Figure 1.1). In general, the abnormal area of the brain will be evident on MRI or other imaging modality. Another example of a symptomatic epilepsy would be one arising from a focal cortical dysplasia. However, some epilepsies are caused not by a clear anatomic abnormality but are instead inherited—either as a single mutation or, more

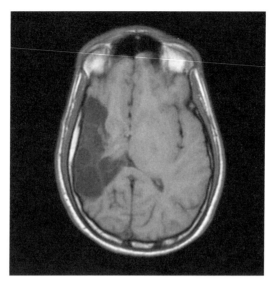

(A)

FIGURE 1.1 MRI images and EEG data from a teenager with symptomatic localization-related epilepsy. This 15-year-old right-handed boy had left-sided hemiplegic cerebral palsy and startle-induced complex partial seizures. Panels A and B are representative T1-weighted postcontrast MRI images (coronal and axial planes, respectively) demonstrating the damage caused by an in utero right middle cerebral artery territory infarction (the left side of the image corresponds to the right side of the brain). Panel C shows the patient's EEG immediately prior to and following an auditory startle as well as several seconds into his typical electrographic ictal discharge. Note the high-amplitude slow activity with superimposed faster frequencies in the leads labeled Fp2-F4, F4-C4, and C4-P4, indicating that the seizure is arising from the right frontocentral region (by convention, EEG leads with even numbers are on the right, and those with odd numbers are on the left; Fp = Frontal polar, F = frontal, C = central, and P = parietal). The patient underwent definitive surgical resection and remains seizure free off medication after more than 4 years.

commonly, as a collection of interacting mutations. Epileptologists refer to such epilepsies as *idiopathic*. Several common epilepsies of childhood, such as childhood absence epilepsy and benign Rolandic epilepsy, are considered idiopathic. Finally, some epilepsies occur in patients without an evident cause: the MRI is normal, there is no clear heritability, and no aspect of the workup reveals a potential etiology. These epilepsies are labeled *cryptogenic*—literally meaning that the cause is hidden. One of the primary goals of the epilepsy research community is to abolish the need for this category by increasing our understanding of what causes epilepsy as well as expanding our repertoire of tools available for diagnosis.

In addition to etiology, the present classification scheme utilizes seizure type as a criterion. Seizures that arise from a particular region of the brain are labeled *partial*, whereas seizures that involve both hemispheres from onset are referred to as *generalized*. Rather than using the term "partial" when categorizing the epilepsies, the ILAE scheme has adopted the term *localization related*. It should be noted

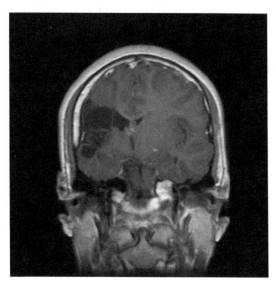

(B)

FIGURE 1.1   (continued)

that seizures may begin focally (as a partial seizure) and then spread to involve the other hemisphere. Such a seizure is said to be *secondarily generalized*. Although not utilized in the classification scheme, partial seizures are further divided into *simple partial seizures* if they do not affect consciousness, and *complex partial seizures* if consciousness is in any manner impaired. Putting the etiologic and phenomenological criteria together yields the appropriate classification. For example, epilepsies may be *symptomatic localization related* (a clear anatomic cause affecting just one part of the brain), *idiopathic generalized* (an inherited epilepsy producing seizures that affect both hemispheres at the outset, such as childhood absence epilepsy), *cryptogenic localization related* (epilepsy with no clear etiology, which arises from a restricted focus), or any other combination of terms. As will become clear in the chapters to follow, utilizing this scheme is helpful in determining an appropriate evaluation and management strategy.

Another peculiarity of pediatric epilepsy is the concept of an epilepsy syndrome: a constellation of a particular type of seizure (or seizures), EEG features, and other clinical phenomenon often associated with a particular age of onset. For example, West syndrome represents the combination of infantile spasms (a particular type of seizure), an interictal EEG pattern called hypsarrhythmia, and developmental arrest or regression with a peak age of onset between 3 and 7 months of age.[15] Although still clinically diverse, epilepsy syndromes seem to represent a more homogeneous clinical population than is afforded by the ILAE classification scheme. For example, childhood absence and juvenile myoclonic epilepsy are both categorized as idiopathic generalized epilepsies, but they differ significantly in their age of onset, predominant seizure type, and rate of remission.

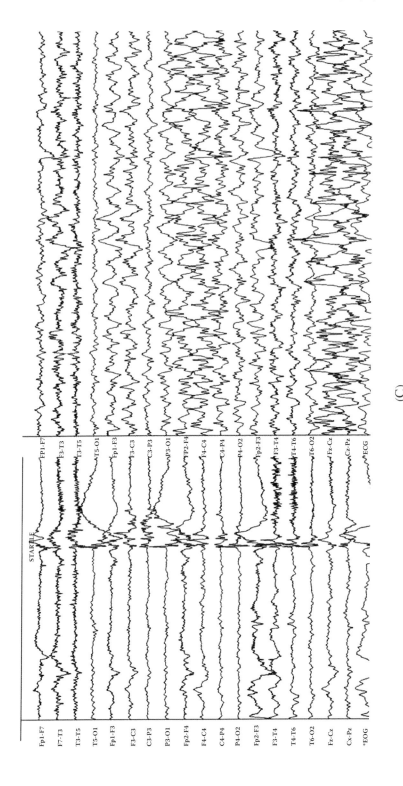

(C)

FIGURE 1.1   (continued)

PEDIATRIC EPILEPSY VIEWED FROM A DEVELOPMENTAL CONTEXT

The CNS is unique in that its development extends from early embryonic life, throughout childhood, and even into early adulthood. This has implications for both the causes and consequences of pediatric epilepsy as well as its treatment. An important determinant of the effects of a developmental insult is the ontogenetic stage at which it occurs. For example, failure of the anterior neuropore to close in the fourth embryonic week would cause anencephaly, whereas an insult in the second trimester might cause a focal cortical dysplasia.[21] For reasons that remain incompletely understood, the immature nervous system seems to be uniquely susceptible to developing seizures. Another way to state this is that the "seizure threshold" of the developing nervous system seems to be lower than that of the adult nervous system. At least part of this susceptibility may be secondary to the ongoing ontogenetic processes of the immature brain.[3,24]

One possible contributor to the decreased seizure threshold of the immature nervous system is a physiological imbalance of excitation and inhibition. In general, excitatory synaptic connections develop before inhibitory ones. Further, very early in development, it appears that inhibitory neurotransmission is actually depolarizing and therefore possibly excitatory. This appears to be due to the developmental expression of a particular type of cation chloride cotransporter that produces a more positive (depolarized) chloride reversal potential than what is found in the mature nervous system.[7] This is a potentially clinically relevant physiological phenomenon because most first-line anticonvulsants used to treat neonatal seizures—barbiturates and benzodiazepines—act by increasing inhibition. Maturation of the GABA-ergic system also involves expression of different receptor isoforms as well as unique modulatory neuropeptides (such as somatostatin). Overall, relatively late emergence of functional inhibition may increase the propensity of the CNS toward excessive excitation, which increases the likelihood of seizures.

The process of synaptogenesis involves abundant synapse formation followed by activity-dependent pruning of ineffective, aberrant, or unnecessary connections. Such developmental plasticity requires the developing nervous system to be uniquely responsive to environmental effects. Because of this, insults can have pervasive and persistent effects. Excessive activity during critical periods of development may strengthen neuronal pathways that subsequently form a seizure focus or become pathways of seizure propagation. Indeed, this may be one important component of the process of developmental epileptogenesis. Relative immaturity of cortical connections is also important for the clinical appearance of seizures. For example, neonates, who physiologically lack extensive well-formed intercortical and interhemispheric connectivity, do not exhibit generalized seizures.

Formation of the cerebral cortex is an intricate and remarkable process that begins with cells becoming neurons near the ventricles, followed by migration of these new neurons to their appropriate location in the cortex.[16] Interestingly, this process proceeds in an "inside-out" fashion—with the most recently generated neurons migrating through cells forming the more inner cortical layers. This choreographed relocation of cells involves glial cells, called radial glia, upon which the neurons migrate, as well as morphological cues to guide their entrance to and exit from this pathway. As might be expected, given the inherent complexity of this process, not all cells successfully reach their designated location. Such "heterotopic" neurons are likely of little consequence if found in isolation, as they are a common incidental finding in the brains of normal individuals without epilepsy. However, in some patients, a collection of neurons fails to completely migrate and may become a focal cortical dysplasia.[20] The extent of dysplastic cortex can range from quite restricted to very extensive. For reasons that are incompletely understood, such foci of an abnormally formed cortex are often highly epileptogenic and are commonly found in children with localization-related epilepsy.

In addition to the developmental causes of epilepsy, clinicians caring for children with epilepsy must always be mindful of the potential developmental consequences of our treatments.[13] Antiepileptic medications, in general, act by increasing inhibition or decreasing excitation. Such therapeutic manipulations interact with the ongoing process of synaptogenesis and may alter cognitive processes. This is one important reason to be judicious in the use of medical therapy because it can carry its own set of potential morbidities.

A WIDE RANGE OF TREATMENT OPTIONS IS AVAILABLE

Once the diagnosis of epilepsy has been made, consideration turns to appropriate treatment. Some forms of childhood epilepsy may not require any intervention other than education and reassurance. For example, benign Rolandic epilepsy is a common idiopathic localization-related epilepsy of childhood that spontaneously resolves by the age of 20.[8] Benign Rolandic epilepsy is also referred to as "benign epilepsy with centro-temporal spikes" (BECTS). Approximately 60% of patients with this condition experience very few seizures. When seizures do occur under such circumstances, an abortive therapeutic option—such as a rectally administered form of diazepam—is often prescribed in lieu of daily medical therapy. For those children with recurrent unprovoked seizures that are sufficiently frequent to require intervention, there are at least 16 different medications from which to choose. Factors such as type of epilepsy, age of the patient, and comorbidities are important considerations in deciding which medication to use. Perhaps the single most important factor in medication choice is the specific side-effect profile of the drug and its suitability for a particular patient. Overall, approximately 60% of patients will become seizure free with one of the first two anticonvulsants prescribed.[17] Unfortunately, for those whose epilepsy does not respond, the chance of treatment success with subsequent medication trials is much less. For this reason, patients who do not respond to one of the first two or three medications are referred to as "medically refractory."

Fortunately, there is an ever-increasing range of options for patients with medically refractory epilepsy. One possibility is the use of the ketogenic diet: a high-fat and low-carbohydrate protocol, which results in ketone body production and which produces improved seizure control.[10] For certain carefully selected patients, the best option is epilepsy surgery: for example, neurosurgical removal of the focus of the epileptogenic cortex. Examples of such procedures range from focal neocortical resection for patients with an area of cortical dysplasia to removal of the anterior temporal lobe in patients with temporal lobe epilepsy to hemispherectomy in patients with hemimegalencephaly. Epilepsy surgery candidates undergo an extensive presurgical evaluation that includes neuroimaging, EEG monitoring, and detailed neuropsychological studies, as well as other ancillary tests. Given the irreversible nature of surgical intervention, it is vital to determine whether potential functional deficits might result from the proposed resection. Still, for excellent candidates, the chance of becoming seizure free following surgery is as high as 65–70%, depending principally on the location of the focus and whether or not there exist clear imaging findings related to that focus.[25] At times, surgical procedures are conducted with the goal of decreasing seizure frequency or for palliation. This may involve, for example, partial resection of a lesion if the presence of the eloquent cortex prevents complete removal, or corpus callosotomy to prevent secondary generalization of seizures.

Another surgical option used to decrease seizure frequency is implantation of a vagus nerve stimulator (VNS).[19] This device consists of a generator implanted subcutaneously over the pectoral muscle, which is connected via leads wrapped around the left vagus nerve. The VNS has an adjustable stimulation cycle that delivers pulses of defined intensity and duration to the vagus nerve. For unknown reasons, such stimulation significantly decreases seizure frequency in approximately 30% of patients. Although unlikely to make a patient seizure free, it may significantly improve seizure control. Presently in clinical trials is a unique responsive neurostimulator that utilizes focal cortical stimulation to abort partial seizures. As the range of therapeutic interventions for medically refractory epilepsy expands, it becomes ever more vital to refer such patients to an epilepsy center where the possible use of such therapies can be considered.

PEDIATRIC EPILEPSY IS NOT JUST ABOUT SEIZURES

Although seizures are surely the most dramatic aspect of epileptic disorders, they are far from the only clinical manifestation. Compared to children with other chronic medical conditions, patients with epilepsy have a lower rate of successful educational completion, employment, marriage, and other important quality-of-life measures.[14] Rates of affective disorders and behavioral problems are also much higher than in the general population.[2] Interestingly, this remains true even for patients with epilepsy that readily comes under medical control, as well as patients who undergo successful epilepsy surgery. Certainly many patients with epilepsy do extremely well, yet it remains troubling that there are those who do not. For some, frequent seizures can result in an encephalopathy that interferes with psychosocial function. Also, because anticonvulsant medications work by increasing inhibition or decreasing excitation, cognitive dysfunction is a frequent side effect. Still another aspect of this multifactorial phenomenon is that epilepsy reflects, at some level, neuronal dysfunction. Therefore, it is perhaps not surprising that patients with epilepsy may also have difficulties with other cortically and subcortically mediated processes. This possibility is further suggested by the finding that behavioral problems in children with epilepsy precede the diagnosis of epilepsy approximately 25% of the time.[1] Regardless of the underlying pathophysiology, it is crucial that we consider such comorbidities in caring for our patients with epilepsy.

SUMMARY

Given the relatively high incidence of epilepsy, all physicians who work with children, regardless of specialty, will encounter patients afflicted with this heterogeneous disorder. Appropriate care of these patients is crucial given the potential developmental consequences of both the underlying epileptogenic process and the treatments we employ. The therapeutic armamentarium available to epilepsy clinicians continues to expand as does our understanding of the basic neurobiology of these conditions. Yet, as we work with our patients to make them seizure free, we must also be continually cognizant of the wide-ranging effects of the epilepsies and avoid focusing simply on the seizures themselves.

REFERENCES

1. Austin J., Harezlak J., Dunn D., Huster G., Rose D., Ambrosius W. (2001) Behavior problems in children before first recognized seizures. *Pediatrics* 107(1): 115–122.
2. Baker G. (2006) Depression and suicide in adolescents with epilepsy. *Neurology* 66: S5–S12.
3. Bender R., Baram T. (2007) Epileptogenesis in the developing brain: what can we learn from animal models? *Epilepsia* 48(Suppl. 5): 2–6.
4. Commission on Classification and Terminology of the International League Against Epilepsy. (1989) Proposal for revised classification of epilepsies and epileptic syndromes. *Epilepsia* 30: 389–399.
5. Eder W., Ege M., Mutius, E. (2006) The asthma epidemic. *N. Engl. J. Med.* 355: 2226–2235.
6. Engel J. (1999) A proposed diagnostic scheme for people with epileptic seizures and with epilepsy: report of the ILAE Task Force on Classification and Terminology. *Epilepsia* 42: 796–803.
7. Galanopoulou A. (2007) Developmental patterns in the regulation of chloride homeostasis and GABA$_A$ receptor signaling by seizures. *Epilepsia* 48(Suppl. 5): 14–18.
8. Gobbi G., Boni A., Fillippini M. (2006) The spectrum of idiopathic Rolandic epilepsy syndromes and idiopathic occipital epilepsies: from the benign to the disabling. *Epilepsia* 47(Suppl 2): 62–66.
9. Guerrini, R. (2006) Epilepsy in children. *Lancet.* 367: 499–524.
10. Hartman A., Gasior M., Vining E., Rogawski M. (2007) The neuropharmacology of the ketogenic diet. *Pediatr. Neurol.* 36(5): 281–292.
11. Hauser W., Annegers J., Kurland L. (1993) Incidence of epilepsy and unprovoked seizures in Rochester, Minnesota: 1935–1984. *Epilepsia* 34(3): 453–468.
12. Kho L., Lawn N., Dunne J., Linto J. (2006) First seizure presentation: Do multiple seizures within 24 hours predict recurrence? *Neurology* 67: 1047–1049.
13. Kim J., Kondratyev A., Tomita Y., Gale K. (2007) Neurodevelopmental impact of antiepileptic drugs and seizures in the immature brain. *Epilepsia* 48(Suppl. 5): 19–26.
14. Kobau R., Zahran H., Grant D., Thurman D., Price P., Zack M. (2007) Prevalence of active epilepsy and health-related quality of life among adults with self-reported epilepsy in California: The California Health Interview Survey, 2003. *Epilepsia* 48(10): 1904–1913.
15. Korff C., Nordli D. (2006) Epilepsy syndromes in infancy. *Pediatr. Neurol.* 34: 253–263.
16. Kriegstein A., Noctor S. (2004) Patterns of neuronal migration in the embryonic cortex. *Trends Neurosci.* 27(7): 392–399.

17. Kwan P., Brodie M. (2000) Early identification of refractory epilepsy. *N. Engl. J. Med.* 342(5): 314–319.
18. Lizana J., Garcia E., Marina L., Lopez M., Gonzalez M., Hoyos A. (2000) Seizure recurrence alter a first unprovoked seizure in childhood: A prospective study. *Epilepsia* 41(8): 1005–1013.
19. McHugh J., Singh H., Phillips J., Murphy K., Doherty C., Delanty N. (2007) Outcome measurement after vagal nerve stimulation therapy: proposal of a new classification. *Epilepsia* 48(2): 375–378.
20. Najm I., Tilelli C., Oghlakian R. (2007) Pathophysiological mechanisms of focal cortical dysplasia: A critical review of human tissue studies and animal models. *Epilepsia* 48(Suppl. 2): 21–32.
21. Nolte J. (1999) *The Human Brain, 2nd Ed.* St. Louis, MO: Mosby.
22. Pohlmann-Eden B., Beghi E., Camfield C., Camfield P. (2006) The first seizure and its management in adults and children. *BMJ* 332: 339–342.
23. Sadleir L., Scheffer, I. (2007) Febrile seizures. *BMJ* 334: 307–311.
24. Sankar R., Rho J. (2007) Do seizures affect the developing brain? Lessons from the laboratory. *J. Child Neurol.* 22(Suppl): 21S–29S.
25. Tellez-Zenteno J., Dhar R., Wiebe S. (2005) Long-term seizure outcomes following epilepsy surgery: a systematic review and meta-analysis. *Brain* 128(5): 1188–1198.

2 Developmental Pharmacokinetics

Principles and Practice

Gail D. Anderson, Ph.D. and Jong M. Rho, M.D.

CONTENTS

Many age-related variables influence the pharmacokinetic properties of drugs.[37] For example, with respect to absorption, gastric pH is increased in neonates, infants, and young children, but decreases to adult levels after 2 years of age. Gastric and intestinal motility is decreased in neonates and infants, but is increased in older infants and children to comparable adult levels. Very few studies have evaluated the maturation of the rate and extent of absorption; however, it is generally accepted that the absorption rate of drugs is lower in neonates and young infants compared to older children. There is little or no information regarding the maturation of active transporters in the gastrointestinal tract that are known to significantly affect the bioavailability of certain drugs.

Once a drug is absorbed, it is distributed to various body compartments in a manner that is dependent on its unique physiochemical properties, such as molecular size, ionization constant, and relative aqueous and lipid solubility. In neonates and infants, the increased total-body-water to body-fat ratio contributes to an increase in the volume of distribution (V_d) of drugs. The direction of the change (i.e., increase or decrease in V_d) will also depend on the drug's physiochemical characteristics. The plasma concentration that results from a loading dose of a drug is inversely proportional to the V_d. Therefore, determination of loading doses for a given drug should account for age-related changes in V_d. For example, neonates and young infants require larger loading doses of the antiepileptic drugs (AEDs) phenobarbital and valproic acid to attain similar therapeutic plasma concentrations found in adults due to an increased V_d.

Protein binding is another important variable that affects V_d. Albumin and α_1-acid-glycoprotein concentrations are decreased in the neonate and infant, and reach adult levels only by 1 year of life. The decreased protein binding alters the ratio of unbound to total plasma concentrations of the AEDs. For AEDs that are highly protein bound, such as phenytoin, valproic acid, and tiagabine, total concentrations are not reliable for therapeutic drug monitoring and will underestimate the unbound or active concentration of AEDs in neonates. Assessments of unbound plasma concentrations are required to avoid dose-dependent adverse events.

Elimination of AEDs occurs through either renal excretion of unchanged parent drug, hepatic biotransformation to metabolites (both active and inactive), or a combination of both. At birth, renal blood flow, glomerular filtration rates, and tubular secretion and reabsorption are at approximately 25–30% of adult values, but increase steadily by 6 months to 50–75% of adult function. Full maturation of renal function is generally reached by approximately 1 year of age. As with the gastrointestinal tract, transporter proteins participate in active renal excretion of many drugs; however, knowledge regarding their maturation remains scant. In general, weight-normalized doses of drugs, excreted predominately unchanged by the kidneys, need to be reduced only for neonates and infants. The cytochrome P450 (CYP) and uridine diphosphate (UDP) glucuronosyltransferase (UGT) family of enzymes catalyze biotransformation of most of the older AEDs. The more recently approved AEDs are also eliminated by renal, mixed, and non-CYP or non-UGT pathways. Drug interactions occur less frequently with drugs metabolized by non-CYP or non-UGT pathway or if eliminated unchanged.

The influence of age on hepatic metabolism is dependent on the types of enzymes involved. CYP-dependent metabolism is low at birth—approximately 50 to 70% of adult levels. However, by 2 to 3 years of age, CYP enzymatic activity actually exceeds adult values. Therefore, infants less than 1 year of age generally have decreased ability to eliminate drugs, whereas young children have an increased ability (relative to adults) to eliminate drugs metabolized by the CYP isozymes, including CYP1A2, CYP2C9, and CYP3A4. By puberty, the CYP activity decreases to adult levels. The one exception is CYP2C19, which appears to have similar activity in children as that found in adults. UGT activity in neonates is deficient at birth, and reaches adult levels by 2 to 4 years of age. Children appear to have slightly increased UGT activity compared to adult levels; however the differences between activity in children and adults is significantly less than with the CYP isozymes. For drugs metabolized by non-CYP or non-UGT enzymes, the effect of age is unknown. The effect of age on drugs that are eliminated by a combination of pathways (i.e., renal and hepatic) will depend both on the relative maturation of these pathways, as well as the relative fraction of each drug eliminated through them. The AEDs can be divided into those that are eliminated by hepatic metabolism (CYP, UGT, or non-CYP/UGT), exclusively renal excretion of unchanged drug, or a combination of both renal and hepatic elimination.[51] Maturation in both absorption (bioavailability of) and elimination processes (clearance) will affect the relationship between the average steady-state concentrations obtained and the dose (concentration/dose). The pharmacokinetic properties of the AEDs are summarized in Table 2.1.

TABLE 2.1
Pharmacokinetic properties of the antiepileptic drugs

Drug	Absorption		Protein bound	Children require larger mg/kg doses?[a]	How much?	Effect of enzyme inducers?
	Bioavailability (F)	T_{max} (h)				
Renal elimination						
Gabapentin	< 60%	2–3	0	Yes	30–50%	No
Pregabalin	> 90%	3	0	NK[b]		No
Vigabatrin	60–80%	0.5–2	0	No		No
Metabolic elimination						
Carbamazepine	> 80%	4–8	75	Yes	50–100%	Yes
Clobazam	> 85%	0.5–2	90	No		Yes
Clonazepam	> 80%	1–4	85	No		Yes
Diazepam	> 95%	0.5–1.5	94–99	Yes	50–100%	Yes
Lamotrigine	> 95%	2–4	50	No		Yes
Lorazepam	> 90%	1–2	85–91	No		Yes
Phenytoin	> 90%	varied	90	Yes	50–100%	Yes
Tiagabine	>95%	1	96	Yes	50%	Yes
Valproic acid	>95%	varied	7–15	Yes	50–100%	Yes
Metabolic and renal elimination						
Ethosuximide	100%	3–7	0	Yes	50–100%	Yes
Felbamate	> 95%	2–4	30	Yes	40%	Yes
Levetiracetam	> 95%	1	0	Yes	30–50%	No
Oxcarbazepine	> 95%	7 (MHD)[c]	40	Yes	30–80%	Yes
Phenobarbital	> 95%	1–4	50	Yes	50–100%	Yes
Topiramate	~ 80%	3–4	20	Yes	30–50%	Yes
Zonisamide	> 90%	2–4	50	Yes	50–100%	Yes

[a] Approximately estimated based on available data.
[b] Not known.
[c] Oxcarbazepine is a pro-drug. Pharmacokinetics in the data provided is for MHD, the active metabolite (see text).

AEDS ELIMINATED RENALLY

Of AEDs eliminated unchanged by renal excretion (i.e., gabapentin, pregabalin, vigabatrin), neonates and infants should require significantly lower doses than children and adults because of immature renal function. Weight-corrected doses should be approximately the same in children and adults if there are no age-related effects on absorption. Gabapentin is less than completely absorbed (<60%) and is highly variable because of saturation of active L-neutral amino acid transporters in the gastrointestinal tract.[28] A population pharmacokinetics analysis of gabapentin in infants and children (2 months to 13 years) found that children younger than 5 years of age had significantly higher and more variable oral clearance than older children.[48] Infants and young children under 5 years of age required 33% higher weight-normalized doses of gabapentin to attain similar concentrations. The weight-normalized oral clearance in children older than 5 years was comparable to that obtained in adults. As creatinine clearance was not significantly different between younger and older children, the age-related difference might be due to decreased oral bioavailability—possibly caused by delayed maturation of the L-neutral amino acid transporter. Two smaller studies found a 33%[27] and 50%[7] higher oral clearance with children aged 10 years or less, compared to young adults. Therefore, children younger than 5 years will need higher weight-normalized doses than children older than 5 years. Further, there is some evidence that children older than 5 years may also need higher weight-normalized doses than adults. There are currently no data regarding the use of pregabalin in infants and children.

Vigabatrin is eliminated almost completely unchanged by the kidneys.[35] A single-dose study in two groups of 6 children, aged 5 months to 2 years and aged 4 to 14 years, found an age-related effect only for the pharmacological inactive R-(–)-enantiomer. The age-related difference was related to bioavailability and not renal elimination. Armijo et al.[6] evaluated racemic vigabatrin plasma concentrations in a larger population of 65 adults and 114 children. The concentration-to-dose ratio of the racemic vigabatrin was significantly lower in children aged 1 to 9 years, compared to children aged 10 to 14 years. Surprisingly, the ratio was also lower in children aged 10 to 14 years than those older than 15 years. Children younger than 5 years had 50% lower ratios than adults. The mechanism of this age-related difference is not known. The clinical significance is also unclear as there is a lack of a documented relationship between vigabatrin plasma concentrations and clinical effect. The mechanism of action of vigabatrin is irreversible inhibition of the enzyme GABA (γ-aminobutyric acid) transaminase, which results in a biological half-life significantly lower than the plasma elimination half-life. Therefore, vigabatrin is titrated slowly to clinical effect and not to therapeutic plasma concentrations.

AEDS ELIMINATED BY CYP-DEPENDENT METABOLISM

For the AEDs listed in Table 2.1 that are eliminated predominantly by CYP-dependent metabolism (i.e., carbamazepine, diazepam, phenytoin), neonates and infants will require lower doses than young children. Young children will require approximately 50% higher mg/kg doses than older children, and older children will require approximately 50% higher mg/kg doses than adults. Carbamazepine is extensively metabolized, with less than 1% excreted unchanged in the urine. CBZ-epoxide is the predominant metabolite, accounting for approximately 25% of the dose in monotherapy and 50% in polytherapy when other enzyme-inducing AEDs are present. When considering effects of age on carbamazepine, one also must consider the effects on the active metabolite, which is rarely measured in the clinical setting. CBZ-epoxide is pharmacologically active and contributes to the therapeutic effects of carbamazepine as well as to its neurotoxicity. Studies with carbamazepine have found a higher weight-adjusted total body clearance and higher CBZ-epoxide-to-carbamazepine ratio in children compared to adults.[11,53] Adult values are reached by 15 to 17 years of age, with the greatest change in oral clearance occurring between 9 and 13 years of age.[2] The significantly shorter $t_{1/2}$ in children compared to adults may require three times a day (TID) or even four times a day (QID) dosing of the tablet and suspension dosage forms. The use of controlled-release or sustained-release formulations provides significantly less fluctuation in plasma concentrations and decreased toxicity associated with high peak concentrations.[17] In general, children need approximately 50 to 100% higher weight-normalized maintenance doses than adults to achieve comparable serum levels.

Diazepam is extensively metabolized to several active substances, including desmethyldiazepam, temazepam, and oxazepam, through reactions catalyzed by CYP2C19 and CYP3A4. The mean $t_{1/2}$ of diazepam and desmethyldiazepam is significantly prolonged in poor metabolizers of CYP2C19. The recommended dosing of rectal gel diazepam for treatment of acute serial seizures in children does reflect the expected increased weight-normalized oral clearance for drugs metabolized by CYP3A4: ages 2 to 5 years (0.5 mg/kg), children 6 to 11 years (0.3 mg/kg), and children and adults >11 years (0.2 mg/kg).

Phenytoin is eliminated predominately by saturable CYP2C9- and CYP2C19-dependent hepatic metabolism.[18] In both children and adults, carriers of CYP2C9 or CYP2C19 mutant alleles will exhibit significantly increased concentration-to-dose ratios.[40,47,54] Phenytoin is described by a capacity-limited metabolism (i.e., Michaelis–Menten kinetics), where V_{max} is the maximum rate of metabolic capacity. Neonates have a smaller V_{max}, resulting in a decreased weight-normalized unbound clearance. In neonates, decreased albumin results in decreased protein binding and, as such, total phenytoin concentrations will not reflect the unbound or active phenytoin. Children have a significantly higher mean V_{max} than adults, which progressively declines during childhood and reaches adult values around puberty.[16,26]

AEDS ELIMINATED BY UGT-DEPENDENT HEPATIC METABOLISM

For drugs eliminated predominately by UGT (i.e., lamotrigine and lorazepam), neonates and infants will require lower doses; however, weight-corrected doses should be approximately the same for children and adults. Lamotrigine is predominately eliminated by hepatic metabolism as a UGT1A4-catalyzed glucuronide conjugate.[32] In infants under 1 year of age with infantile spasms or partial seizures, oral clearance of concomitant medications (which includes enzyme inducers), increased during the first year of life.[20] During the first month, oral clearance (weight-normalized) was approximately 50% lower than in infants 2 to 12 months of age. There are conflicting data regarding age-related effects on oral clearance of lamotrigine in older children. Some studies have reported a trend toward a decreased concentration-to-dose ratio in children versus adults; however, few children were receiving lamotrigine monotherapy.[5,10,12] Whether age affects induction potential remains unclear. It is possible that the lower concentration-to-dose ratios in children on enzyme-inducing drugs may be due to increased induction capacity. Chen et al.[22] compared the oral clearance in a group of 12 children in the absence of other AEDs after a single dose of lamotrigine. Even though there was a trend toward increased oral clearance in the four children younger than 6 years of age, compared to the eight children aged 6 to 11 years, there was large intersubject variability. Overall, the single-dose oral clearance was not significantly different from the oral clearance found after single-dose lamotrigine in normal adult subjects in several studies.[23,52] Therefore, doses of lamotrigine in children receiving monotherapy should be similar to adult doses; however, with concomitant enzyme-inducing therapy, children may need higher weight-normalized doses.

Lorazepam is extensively metabolized to a glucuronide conjugate with little renal excretion of the unchanged drug. Neonates have significantly decreased oral clearance compared to children and adults.[38,44] There is no information on the pharmacokinetics of lorazepam in children between the ages of 1 and 7 years. In a group of children aged 7 to 19 years, the pharmacokinetics of lorazepam was not significantly different from values obtained in adults. Because glucuronidation appears to reach adult levels by age 2 to 3 years, after 3 years of life, weight-corrected doses in children should be the same as in adults. Infants and young children <3 years should receive reduced doses.

AEDS ELIMINATED BY MIXED CYP, UGT, AND OTHER METABOLIC PATHWAYS

Several of the AEDs (i.e., clobazam, clonazepam, tiagabine, valproic acid) are extensively metabolized by multiple metabolic pathways, with minimal excretion of drug unchanged in the urine. Predicting age-related effects is more difficult because of the larger intersubject variability in the fraction eliminated by each pathway and lack of data on the effect of age on the non-CYP and UGT enzymes. Clobazam is eliminated predominately by hepatic metabolism to multiple metabolites. The primary metabolite, N-desmethylclobazam, is active and accumulates to approximately eight-fold higher serum concentrations than clobazam after multiple dosing. In a population analysis of over 400 epileptic patients receiving different comedications, Bun et al. found that N-desmethylclobazam concentrations were significantly lower in children compared to those in adults. They did not find an age-related difference in concentration-to-dose ratio with clobazam.[19] In another population study of 74 children, Theis et al.[58] noted that both clobazam and the metabolite concentrations increased with increasing age from 1 to 18 years. In both studies, there was very large intersubject variability and a poor correlation between plasma concentrations and therapeutic efficacy. Based on this limited data, it is unclear if children require higher doses of clobazam than adults. Doses of clobazam should be initiated and titrated to effect in both children and adults.

Clonazepam is extensively metabolized to inactive metabolites with less than 1% excreted unchanged in the urine. Neonates receiving clonazepam require lower weight-normalized doses than older children and adults. The elimination half-life ($t_{1/2}$) of clonazepam is prolonged, with a significantly lower clearance than found in older children and adults.[4] In studies involving small numbers of children, the weight-normalized oral clearance was found to be highly variable although not significantly different from that in adults.[25,59]

Tiagabine is extensively metabolized via CYP3A4- and UGT-predominate pathways, with less than 2% excreted unchanged in the urine.[34] The weight-corrected clearance of tiagabine is twofold higher in children than adults not receiving enzyme-inducing AED polytherapy (i.e., valproic acid). In children receiving enzyme-inducing drugs, the weight-corrected clearance is similar to that in adults.[33] At present, there is not enough clinical evidence to suggest that children receiving tiagabine without an enzyme-inducing drug will need 50% higher doses than adults.

Valproic acid undergoes extensive hepatic metabolism, with less than 5% of the dose excreted unchanged in the urine.[39] Major metabolism occurs by UGT-catalyzed glucuronide conjugation and β-oxidation with a minor CYP-dependent component. Neonates with intractable seizures have highly variable but similar total clearances as compared to adults. Due to low albumin concentrations in neonates, total valproic acid concentrations underestimate the unbound or pharmacologically active valproic acid concentration. During the first 2 months of life, clearance increases significantly because of maturation of hepatic enzymes. Older infants, aged 3 to 36 months, had weight-normalized clearance values significantly higher than that found with adults.[36] School-age children have clearances intermediate to those found in infants and adults.[21] Infants and young children need weight-adjusted doses 50–100% higher than adults to attain similar valproic acid plasma concentrations.

AEDS ELIMINATED BY HEPATIC METABOLISM AND RENAL EXCRETION

Many of the currently available AEDs (i.e., ethosuximide, felbamate, levetiracetam, oxcarbazepine, phenobarbital, topiramate and zonisamide) are eliminated by a combination of metabolism and renal excretion. Ethosuximide is eliminated primarily by CYP3A4-dependent metabolism, with approximately 20% excreted unchanged in the urine. The weight-adjusted oral clearance of ethosuximide is higher in children than adults, and the ratio of ethosuximide concentration to dose was found to be 50% higher in children aged 2.5 to 10 years compared to older children (≥15 years).[14] Children need approximately 50–100% higher mg/kg maintenance doses than adults to attain similar ethosuximide concentrations.

Felbamate is eliminated via renal excretion of unchanged drug (50%) and glucuronidation (20%), and is a substrate for CYP3A4 (20%) and CYP2E1. The weight-adjusted apparent clearance of felbamate is approximately 40% higher in children aged 2 to 12 years than adults on monotherapy or polytherapy with other AEDs.[9] There was a significant negative correlation in apparent clearance in a group of 17 children aged 2 to 12 years, with higher clearance in the very young children and decreases to adult values by age 12.[20] Children will require weight-normalized maintenance doses approximately 40% higher than adults to attain similar felbamate concentrations.

Levetiracetam is eliminated predominantly by renal excretion of unchanged drug and by hydrolysis of the acetamide group, a reaction catalyzed by amidases—enzymes that are present in a number of tissues. Weight-normalized oral clearance of levetiracetam in children ages 6 months to 4 years[44] and 6 to 12 years[42,50] is approximately 30 to 40% higher than in adults. Slightly lower clearance rates were found in children aged 2 to 6 months due to immature renal function.[30] Therefore, children older than 6 months will require 30–50% higher weight-normalized doses than adults to achieve similar concentrations.

Oxcarbazepine is a prodrug that is rapidly converted to 10,11-dihydro,10-hydroxycarbazepine monohydroxy-derivative (MHD) upon oral administration, a reaction that is catalyzed by cytosolic arylketone reductase.[41] MHD is predominantly excreted unchanged in the urine or conjugated by UGT and then excreted, with only minor oxidation metabolism to dihydroxy-derivative (DHD). In a study of children aged 2 to 12 years who received a single oral dose of oxcarbazepine, the dose and weight-normalized area under the concentration-time curve (AUC) of MHD was approximately 60% less in children between 6 to 12 years than children 2 to 6 years.[15,49] Similar results were confirmed by other investigators.[8,56] In addition, children aged 6 to 12 years exhibited higher weight-normalized clearance of MHD. Therefore, children under 6 years and between 6 and 11 years of age will require 80% and 30% higher weight-normalized doses, respectively, than adults to achieve similar concentrations.

Phenobarbital is eliminated by both renal excretion of unchanged drug and hepatic metabolism to parahydroxyphenobarbital (a reaction primarily catalyzed by CYP2C9 and CYP2C19) and glucosidation to phenobarbital N-glucoside.[3] Newborns receiving phenobarbital have a decreased clearance compared to young infants and children. During the first year of life, young children have a two- to three-fold greater weight-normalized clearance than adults. Therefore, weight-normalized maintenance doses of phenobarbital in children should generally be 50–100% higher than that in adults.

Topiramate is eliminated as a combination of hepatic metabolism and renal excretion of unchanged drug. Weight-adjusted clearance of topiramate is higher in children aged 4 to 11 years than in adults, resulting in approximately 33% lower topiramate concentrations.[1,13,26,43,55,57] The weight-adjusted clearance of topiramate is slightly higher in infants than in children, and significantly higher than in adults, resulting in an increased dose requirement.[29] With topiramate, titration to effect, and not dose, is recommended in infants and children.[29]

Zonisamide is eliminated by a combination of renal excretion of unchanged drug and hepatic metabolism via hepatic N-acetylation and reduction to 2-sulfamoylacetylphenol (SMAP). Despite considerable experience using zonisamide in children in Japan and Korea, no formal pharmacokinetic studies in children have been completed.[31] In one report, doses of 8 mg/kg/day in 72 children aged 3 months to 15 years resulted in a linear increase in peak and trough concentrations with increasing age.[46] Peak and trough concentrations were approximately two- to three-fold higher in the older children (4–8 µg/mL) compared to infants and young children (2–4 µg/mL). In adults, 200 to 600 mg/day resulted in zonisamide concentrations of 10 µg/mL to 30 µg/mL. Children may require significantly higher doses to achieve zonisamide concentrations comparable to that in adults.

CONCLUSION

It is clear that neonates, young infants, and children undergo significant (and nonlinear) maturational changes in organ systems that prominently affect AED

TABLE 2.2
Age-specific maintenance dosing of antiepileptic drugs used in monotherapy[a]

Drug	Average dose			
	Neonates	Infants	Children	Adults
Phenobarbital	3–4 mg/kg qd	2.5–3.0 mg/kg q12h	2–4 mg/kg q12h	0.5–1.0 mg/kg q12h
Phenytoin	2.5–4.0 mg/kg q12h	2–3 mg/kg q8h	2.3–2.6 mg/kg q8h	2 mg/kg q12h
Carbamazepine	NE	3–10 mg/kg q8h	3–10 mg/kg q8h	5–8 mg/kg q12h
Valproic acid	NE	5–10 mg/kg q8h	5–10 mg/kg q8h	5–10 mg/kg q12h
Ethosuximide	NE	NE	10–20 mg/kg q12h	250–500 mg q12h
Felbamate	NE	NE	5–15 mg/kg q8h	900–1800 mg q12h
Gabapentin	NE	NE	5–15 mg/kg q8h	600–1200 mg q8h
Pregabalin	NE	NE	NE	75–300 mg q12h
Topiramate	NE	NE	2–5 mg/kg q12h	100–200 mg q12h
Lamotrigine	NE	NE	2–5 mg/kg q12h	75–150 mg q12h
Tiagabine	NE	NE	0.5–2 mg/kg qd	32–56 mg qd
Oxcarbazepine	NE	NE	5–15 mg/kg q8h	300–1200 mg q12h
Levetiracetam	NE	NE	5–20 mg/kg q12h	500–1500 mg q12h
Zonisamide	NE	NE	2–6 mg/kg q12h	100–200 mg q12h
Vigabatrin	NE	50–100 mg/kg q12h	25–75 mg/kg q12h	1000–1500 mg q12h

Note: NE = not established.

[a] Not all antiepileptic drugs (AEDs) have FDA-approved indications for monotherapy. When used in conjunction with other AEDs, or drugs that affect hepatic metabolism and/or renal function, doses should be adjusted according to clinical judgment.

absorption, distribution, metabolism, and excretion. The net implication of these changes is that dosing needs to be carefully adjusted if therapeutic serum concentrations are to be achieved, and hence a greater likelihood of achieving seizure freedom (see Table 2.2). In general, for young infants and children, there is a need for increases in weight-normalized dosages of AEDs as these patients have a greater capacity for drug disposition than adolescents and adults. Of course, clinical judgment, combined with judicious use of serum levels and scrutiny of concomitant medications for negative drug interactions, is required to maximize clinical efficacy and tolerability. There currently exist limited data on age-dependent pharmacokinetic properties of AEDs. However, as advances in our understanding of pharmacokinetics, pharmacogenomics, and drug interactions are made, clinicians will hopefully avail themselves of this information to optimize the enduring mainstay of epilepsy therapeutics—antiepileptic drugs.

REFERENCES

1. Adin J, Gomez MC, Blanco Y, Herranz JL, Armijo JA. (2004) Topiramate serum concentration-to-dose ratio: influence of age and concomitant antiepileptic drugs and monitoring implications. *Ther. Drug Monit.* 26(3): 251–7.
2. Albani F, Riva R, Contin M, Baruzzi A. (1992) A within-subject analysis of carbamazepine disposition related to development in children with epilepsy. *Ther. Drug Monit.* 14(6): 457–60.
3. Anderson GD. (2002) Phenobarbital: chemistry, biotransformation and pharmacokinetics. In *Antiepileptic Drugs.* 5th edition Levy RH, Mattson RH, Meldrum BS, Perrucca E, Eds., 496–503, Philadelphia: Lippincott Williams & Wilkins.
4. Andre M, Boutroy MJ, Dubruc C et al. (1986) Clonazepam pharmacokinetics and therapeutic efficacy in neonatal seizures. *Eur. J. Clin. Pharmacol.* 30: 585–9.
5. Armijo JA, Bravo J, Cuadrado A, Herranz JL. (1999) Lamotrigine serum concentration-to-dose ratio: influence of age and concomitant antiepileptic drugs and dosage implications. *Ther. Drug. Monit.* 21: 182–90.
6. Armijo JA, Cuadrado A, Bravo J, Arteaga R. (1997) Vigabatrin serum concentration to dosage ratio: influence of age and associated antiepileptic drugs. *Ther. Drug. Monit.* 19(5): 491–8.
7. Armijo JA, Pena MA, Adin J, Vega-Gil N. (2004) Association between patient age and gabapentin serum concentration-to-dose ratio: a preliminary multivariate analysis. *Ther. Drug. Monit.* 26(6): 633–7.
8. Armijo JA, Vega-Gil N, Shushtarian M, Adin J, Herranz JL. (2005) 10-Hydroxycarbazepine serum concentration-to-oxcarbazepine dose ratio: influence of age and concomitant antiepileptic drugs. *Ther. Drug. Monit.* 27(2): 199–204.
9. Banfield CR, Zhu GR, Jen JF et al. (1996) The effect of age on the apparent clearance of felbamate: a retrospective analysis using nonlinear mixed-effects modeling. *Ther. Drug. Monit.* 18(1): 19–29.
10. Bartoli A, Guerrini R, Belmonte A, Alessandri MG, Gatti G, Perucca E. (1997) The influence of dosage, age and comedication on lamotrigine steady state concentrations in epileptic children: a prospective study with preliminary assessment of correlations with clinical response. *Ther. Drug. Monit.* 19: 252–60.
11. Battino D, Bossi L, Croci D et al. (1980) Carbamazepine plasma levels in children and adults: influence of age, dose, and associated therapy. *Ther. Drug. Monit.* 2(4): 315–22.
12. Battino D, Croci D, Granata T, Mamoli D, Messina S, Perucca E. (2001) Single-dose pharmacokinetics of lamotrigine in children: influence of age and antiepileptic comedication. *Ther. Drug. Monit.* 23: 217–22.
13. Battino D, Croci D, Rossini A, Messina S, Mamoli D, Perucca E. (2005) Topiramate pharmacokinetics in children and adults with epilepsy: a case-matched comparison based on therapeutic drug monitoring data. *Clin. Pharmacokinet.* 44(4): 407–16.
14. Battino D, Cusi C, Franceschetti S, Moise A, Spina S, Avanzini G. (1982) Ethosuximide plasma concentrations: influence of age and associated concomitant therapy. *Clin. Pharmacokinet.* 7(2): 176–80.
15. Battino D, Estienne M, Avanzini G. (1995) Clinical pharmacokinetics of antiepileptic drugs in paediatric patients. Part II. Phenytoin, carbamazepine, sulthiame, lamotrigine, vigabatrin, oxcarbazepine and felbamate. *Clin. Pharmacokinet.* 29: 341–69.

16. Bauer LA, Blouin RA. (1983) Phenytoin Michaelis-Menten pharmacokinetics in Caucasian paediatric patients. *Clin. Pharmacokinet.* 8(6): 545–9.
17. Bialer M. (1992) Pharmacokinetic evaluation of sustained release formulations of antiepileptic drugs: Clinical implications. *Clin. Pharmacokinet.* 22(1): 11–21.
18. Brown TR, Leduc B. (2002) Phenytoin: Chemistry and biotransformation. In *Antiepileptic Drugs. 5th ed.,* Ed. Levy RH, Mattson RH, Meldrum BS, Perrucca E, 565–80. Philadelphia: Lippincott Williams & Wilkins.
19. Bun H, Monjanel-Mouterde S, Noel F, Durand A, Cano JP. (1990) Effects of age and antiepileptic drugs on plasma levels and kinetics of clobazam and N-desmethylclobazam. *Pharm. Toxicol.* 67: 136–40.
20. Carmant L, Holmes GL, Sawyer S, Rifai N, Anderson J, Mikati MA. (1994) Efficacy of felbamate in therapy for partial epilepsy in children. *J. Pediatr.* 125(3): 481–6.
21. Cloyd JC, Kriel RL, Fischer JH, Sawchuck RJ, Eggerth RM. (1983) Pharmacokinetics of valproic acid in children: I. Multiple antiepileptic drug therapy. *Neurology* 33: 185–91.
22. Chen C, Casale EJ, Duncan B, Culverhouse EH, Gilman J. (1999) Pharmacokinetics of lamotrigine in children in the absence of other antiepileptic drugs. *Pharmacotherapy* 19: 437–41.
23. Cohen AF, Land GS, Breimer DD, Yuen WC, Winton C, Peck AW. (1987) Lamotrigine, a new anticonvulsant: pharmacokinetics in normal humans. *Clin. Pharmacol. Ther.* 42: 535–41.
24. Dodson WE. (1982) Nonlinear kinetics of phenytoin in children. *Neurology* 32(1): 42–8.
25. Dreifuss FE, Penry JK, Rose SW, Kupferberg HJ, Dyken P, Sato S. (1975) Serum clonazepam concentrations in children with absence seizures. *Neurology* 25: 255–8.
26. Ferrari AR, Guerrini R, Gatti G, Alessandri MG, Bonanni P, Perucca E. (2003) Influence of dosage, age, and co-medication on plasma topiramate concentrations in children and adults with severe epilepsy and preliminary observations on correlations with clinical response. *Ther. Drug. Monit.* 25(6): 700–8.
27. Gatti G, Ferrari AR, Guerrini R, Bonanni P, Bonomi I, Perucca E. (2003) Plasma gabapentin concentrations in children with epilepsy: influence of age, relationship with dosage, and preliminary observations on correlation with clinical response. *Ther. Drug. Monit.* 25(1): 54–60.
28. Gidal BE, Radulovic LL, Kruger S, Rutecki P, Pitterle M, Bockbrader HN. (2000) Inter- and intra-subject variability in gabapentin absorption and absolute bioavailability. *Epilepsy Res.* 40: 123–7.
29. Glauser TA, Miles MV, Tang P, Clark P, McGee K, Doose DR. (1999) Topiramate pharmacokinetics in infants. *Epilepsia* 40: 788–91.
30. Glauser TA, Mitchell WG, Weinstock A et al. (2007) Pharmacokinetics of levetiracetam in infants and young children with epilepsy. *Epilepsia* 48(6): 1117–22.
31. Glauser TA, Pellock JM. (2002) Zonisamide in pediatric epilepsy: review of the Japanese experience. *J. Child. Neurol.* 17(2): 87–96.
32. Green MD, Bishop WP, Tephley TR. (1995) Expressed human UGT1.4 protein catalyzes the formation of quaternary ammonium-linked glucuronides. *Drug Metab. Dispos.* 23: 299–302.
33. Gustavson LE, Boellner SW, Granneman GR et al. (1997) A single-dose study to define tiagabine pharmacokinetics in pediatric patients with complex partial seizures. *Neurology* 48(4): 1032–7.
34. Gustavson LE, Mengel HB. (1995) Pharmacokinetics of tiagabine, a gamma-aminobutyric acid-uptake inhibitor, in healthy subjects after single and multiple doses. *Epilepsia* 36: 605–11.

35. Haegele KD, Schechter PJ. (1986) Kinetics of the enantiomers of vigabatrin after an oral dose of the racemate or the active S-enantiomer. *Clin. Pharmacol. Ther.* 40(5): 581–6.
36. Hall K, Otten N, Johnston B, Irvine-Meek J, Leroux M, Seshia S. (1985) A multivariable analysis of factors governing the steady-state pharmacokinetics of valproic acid in 52 young epileptics. *J. Clin. Pharmacol.* 25(4): 261–8.
37. Kearns GL, Abdel-Rahman SM, Alander SW, Blowey DL, Leeder JS, Kauffman RE. (2003) Developmental pharmacology—drug disposition, action, and therapy in infants and children. *N. Engl. J. Med.* 349(12): 1157–67.
38. Kearns GL, Mallory GBJ, Crom WR, Evans WE. (1990) Enhanced hepatic drug clearance in patients with cystic fibrosis. *J. Pediatr.* 117: 972–9.
39. Levy RH, Shen DD, Abbott FS, Riggs W, Hachad H. (2002) Valproic acid: Chemistry, biotransformation and pharmacokinetics. In Levy RH, Mattson RH, Meldrum BS, Perrucca E, eds., 780–800. Philadelphia: Lippincott Williams & Wilkins.
40. Mamiya K, Ieiri I, Shimamoto J et al. (1998) The effects of genetic polymorphisms of CYP2C9 and CYP2C19 on phenytoin metabolism in Japanese adult patients with epilepsy: studies in stereoselective hydroxylation and population pharmacokinetics. *Epilepsia* 39(12): 1317–23.
41. May TW, Korn-Merker E, Rambeck B. (2003) Clinical pharmacokinetics of oxcarbazepine. *Clin. Pharmacokinet.* 42(12): 1023–42.
42. May TW, Rambeck B, Jurgens U. (2003) Serum concentrations of Levetiracetam in epileptic patients: the influence of dose and co-medication. *Ther. Drug. Monit.* 25(6): 690–9.
43. May TW, Rambeck B, Jurgens U. (2002) Serum concentrations of topiramate in patients with epilepsy: influence of dose, age, and comedication. *Ther. Drug. Monit.* 24(3): 366–74.
44. McDermott CA, Kowalczyk AL, Schnitzler ER, Mangurten HH, Rodvold KA, Metrick S. (1992) Pharmacokinetics of lorazepam in critically ill neonates with seizures. *J. Pediatr.* 120: 479–83.
45. Mikati MA, Fayad M, Koleilat M et al. (2002) Efficacy, tolerability, and kinetics of lamotrigine in infants. *J. Pediatr.* 141(1): 31–5.
46. Miura H. (2000) Developmental and therapeutic pharmacology of antiepileptic drugs. *Epilepsia* 41 Suppl. 9: 2–6.
47. Odani A, Hashimoto Y, Otsuki Y et al. (1997) Genetic polymorphism of the CYP2C subfamily and its effect on the pharmacokinetics of phenytoin in Japanese patients with epilepsy. *Clin. Pharmacol. Ther.* 62(3): 287–92.
48. Ouellet D, Bockbrader HN, Wesche DL, Shapiro DY, Garofalo E. (2001) Population pharmacokinetics of gabapentin in infants and children. *Epilepsy Res.* 47(3): 229–41.
49. Pariente-Khayat A, Tran A, Vauzelle-Kervroedan F et al. (1994) Pharmacokinetics of oxcarbazepine as add-on therapy in epileptic children (abstract). *Epilepsia* 35 (Suppl. 8): 119.
50. Pellock JM, Glauser TA, Bebin EM et al. (2001) Pharmacokinetic study of levetiracetam in children. *Epilepsia* 42(12): 1574–9.
51. Perucca E. (2006) Clinical pharmacokinetics of new-generation antiepileptic drugs at the extremes of age. *Clin. Pharmacokinet* 45(4): 351–63.
52. Posner J, Holdich T, Crome P. (1991) Comparison of lamotrigine pharmacokinetics in young and elderly healthy volunteers. *J. Pharm. Med.* 1: 121–8.
53. Pynnonen S, Sillanpaa M, Frey H, Iisalo E. (1977) Carbamazepine and its 10,11-epoxide in children and adults with epilepsy. *Eur. J. Clin. Pharmacol.* 11(2): 129–33.
54. Rettie AE, Haining RL, Bajpai M, Levy RH. (1999) A common genetic basis for idiosyncratic toxicity of warfarin and phenytoin. *Epilepsy Res.* 35(3): 253–5.

55. Rosenfeld WE, Doose DR, Walker SA, Baldassarre JS, Reife RA. (1999) A study of topiramate pharmacokinetics and tolerability in children with epilepsy. *Pediatr. Neurol.* 20: 339–44.

56. Sallas WM, Milosavljev S, D'Souza J, Hossain M. (2003) Pharmacokinetic drug interactions in children taking oxcarbazepine. *Clin. Pharmacol. Ther.* 74(2): 138–49.

57. Schwabe MJ, Wheless JW. (2001) Clinical experience with topiramate dosing and serum levels in children 12 years or under with epilepsy. *J. Child Neurol.* 16(11): 806–8.

58. Theis JGW, Koren G, Daneman R et al. (1997) Interactions of clobazam with conventional antiepileptics in children. *J. Child Neurol.* 12: 208–13.

59. Walson PD, Edge JH. (1996) Clonazepam disposition in pediatric patients. *Ther. Drug. Monit.* 18: 1–5.

3 Dietary Therapies for Epilepsy

Eric H. Kossoff, M.D.

CONTENTS

WHAT IS THE KETOGENIC DIET?

The ketogenic diet (KD) is a high-fat, adequate protein, very low carbohydrate diet that is both calorie and fluid restricted and carefully calculated by a ketogenic diet-trained dietitian to create and maintain ketosis.[2] Typical foods eaten include butter, eggs, cheese, heavy whipping cream, canola and olive oils, mayonnaise, green vegetables, chicken, hot dogs, and ground beef. Sugar-free and carbohydrate-free snacks can be incorporated to help make the KD more palatable.

Although the KD's exact mechanism of action remains somewhat unclear, and is probably multifactorial, in nearly 100 retrospective and prospective studies since its introduction in 1921, it has been clearly demonstrated as effective.[5,19] The KD causes a >50% reduction in seizures in approximately 55–60% of the children who begin it by 6 months, and seizure reduction is often maintained long term.[4,5] Despite media-reported stories of cases of dramatic responses, seizure freedom only occurs in 10–15% of children. However, considering the very intractable epilepsy these children have at the time of KD onset, many epileptologists believe that the diet is more likely to be effective than an additional antiepileptic trial after 3–4 antiepileptic drugs have been unsuccessful, especially for generalized epilepsies.

The KD is started gradually in a child as an inpatient, typically following a 24- to 48-hour fasting period designed to rapidly induce ketosis.[2] It is provided in a typically 4:1 or 3:1 ratio (of fat to protein and carbohydrate combined) and is slightly calorie and fluid restricted. The fasting has been demonstrated in several retrospective and prospective studies to be unnecessary for long-term control; however, it

29

appears to lead to a more rapid seizure improvement in many children.[1,8] During the 4- to 5-day-long admission period, families and children are educated for several hours per day regarding the KD attributes and its outpatient management. Most centers still believe that the admission period is crucial for education and subsequent long-term diet adherence.

ARE THERE INDICATIONS FOR THE DIET?

Although prior to just 10 years ago there did not appear to be any particular epilepsy etiologies or syndromes more (or less) likely to respond to the KD, this does not appear to be the case now in 2008. The typical child started on the diet is 3–10 years old, with a mixed epilepsy syndrome such as Lennox–Gastaut syndrome. However, recent evidence exists that has demonstrated its efficacy in infants, adolescents, and even adults.[13,16]

Over the past decade, there has been a significant increase in case reports describing seizure reductions higher than those usually seen with the KD for conditions such as severe myoclonic epilepsy of infancy (Dravet syndrome), tuberous sclerosis complex, Rett syndrome, and myoclonic–astatic epilepsy (Doose syndrome).[3] Children with GLUT-1 deficiency and pyruvate dehydrogenase deficiency should be empirically placed on the KD for its metabolic benefits.[3] In addition, infantile spasm appears in several studies to respond especially well to the KD, particularly if the KD is used earlier in the course of the disorder.[11] Children receiving formula-only diets (e.g., infants on formulas and children with gastrostomy tubes) also do extremely well, with one study reporting a two-fold increase in the likelihood of a >90% seizure reduction.[10]

It is also important to recognize that there are metabolic disorders that are contraindications to the KD. Such disorders involve difficulties with the metabolism of a high-fat diet and include pyruvate carboxylase deficiency, carnitine deficiencies, and fatty-acid oxidation defects. Mitochondrial disorders are also a relative contraindication to the diet, although recent literature has suggested that the diet can be successfully maintained in even these patients.[6] Although it is not a true contraindication, children with Lafora body disease do not appear to respond well to the ketogenic diet, and in our anecdotal experience, the same is true of other progressive myoclonic epilepsies. Lastly, children who are candidates for surgery (e.g., a focal dysplasia or stroke) do not appear likely to be seizure free with the KD when compared to resective surgery.[20]

MEDICATIONS AND THE DIET

Considering the intractable nature of the epilepsy of most children starting the KD, it is not surprising that they remain on medications during their time on the KD. For the majority of children, they are not mutually exclusive therapies. Medications are often changed from solution to tablet formulations to ensure an absence of carbohydrates, although some antiepileptic liquid preparations such as levetiracetam, felbamate, and gabapentin do have lower amounts of carbohydrates when provided as liquids.[2] It is important for the physician to realize, however, that the second most common reason for starting the ketogenic diet following seizure reduction is medi-cation reduction. Although our center (John M. Freeman Pediatric Epilepsy Center at the Johns Hopkins Hospital) generally discourages making two changes at once by immediately reducing medications, evidence would suggest it is safe to do so if parents request and physicians believe it is prudent. Phenobarbital and clonazepam have been associated with a slightly higher risk of increased seizures during their withdrawal in children on the diet.

Most antiepileptics do not have level fluctuations when the KD is started. Carbonic anhydrase inhibitors such as topiramate and zonisamide have inherently increased risks of acidosis and kidney stones. When used in combination with the KD, the risk of kidney stones are not increased above that of the diet alone, but acidosis may be. Valproic acid, which has been reported to lower serum carnitine levels in a manner similar to the KD, does not lead to increased side effects when used in combination.

In preliminary research, it appears that zonisamide may be more likely (and con-versely, phenobarbital less likely) to result in seizure reduction when used with the KD by 3 months. Concurrent vagus nerve stimulation (VNS) use has also been iden-tified as potentially synergistic in a multicenter study.[12] Benefits were often immedi-ate and were more likely to occur with regular VNS duty cycles as opposed to rapid cycling settings.

ADVERSE EFFECTS OF THE KETOGENIC DIET

The KD is neither all-natural nor holistic; side effects do occur.[21] Fortunately, they tend to be transient, treatable, and only very rarely lead to KD discontinuation. They are traditionally divided into "common" (>50%), "occasional" (5%), and "rare" (<1%) when evaluated in the literature, and are listed in Table 3.1. The most common adverse effects are a lack of weight gain (or rarely weight loss), constipation, low-grade acidosis, and hypoglycemia during the fasting period. All of these side effects can be easily treated with additional calories, oral fiber products (Miralax™), extra fluids, oral alkalinization (Polycitra K™), and extra glucose in the form of orange juice if symptomatic, respectively.

Less common side effects include kidney stones, dyslipidemia, and diminished growth. Early studies suggested that the risk of either uric acid or calcium carbonate kidney stones was 6%, and was associated with hypercalciuria (urine calcium/creati-nine ratio higher than 0.2). A recent retrospective study has demonstrated that the use of oral alkalinization (Polycitra K™, 2 Meq/kg per day divided twice daily), histori-cally started in the setting of hypercalciuria, is associated with a three-fold decrease in kidney stone risk when used.[18] As a result of this study, since January 2006 our center now empirically starts all children on Polycitra K™ at diet onset. Cholesterol increases by approximately 30% after 6 months on the KD, and then plateaus.[15] Evi-dence also suggests total cholesterol may decrease after several years to near normal levels. Growth is adversely affected by the KD, more so in young infants and after several years on the diet. Increasing protein and calories may improve this particular side effect, if present. Gastrointestinal upset has also been described in large series. Again, none of the previously mentioned side effects typically necessitate KD dis-continuation, and all these can be treated with either KD modifications (usually to

TABLE 3.1

Reported side effects of the ketogenic diet

Common

    Lack of weight gain

    Constipation

    Hypoglycemia (with fasting)

Occasional

    Gastrointestinal upset or gastroesophageal reflux

    Dehydration or acidosis (more frequent with illness)

    Dyslipidemia

    Kidney stones

    Growth retardation (especially in infants)

    Skeletal fractures (more common with long-term use)

Rare (case reports)

    Pancreatitis

    Cardiomyopathy

    Prolonged QT syndrome

    Basal ganglia changes

    Vitamin or mineral deficiencies (if unsupplemented)

    Carnitine deficiency (symptomatic)

the ketogenic ratio) or supplemental medications. Rare side effects reported include vitamin and mineral deficiencies (prevented with typical supplementation), selenium deficiency, pancreatitis, cardiomyopathy, bruising, basal ganglia change, and prolonged QT intervals.

"ALTERNATIVE" KETOGENIC DIETS

MODIFIED ATKINS DIET

Since first reported in 2003, the modified Atkins diet has emerged as a viable dietary treatment for seizures.[7] This diet is less restrictive than, but perhaps equally effective as, the traditional KD and figures in seven papers published to date. The term "modified" describes the lower carbohydrate limit compared to published Atkins diet recommendations for the "induction phase" (10 versus 20 g per day) and the emphasis on high fat intake. The modified Atkins diet is able to induce ketosis without any protein, fluid, or calorie restriction. In addition, this diet does not require an admission or a fast. In reviewing food records of children on this diet, it approximates a 1:1 ratio of fat:carbohydrate and protein, compared to a typical 3:1 or 4:1 ketogenic diet.[14] Children do not appear to reduce their calorie intake while on this diet. Low carbohydrate foods and meals can also be eaten in restaurants.

TABLE 3.2
Modified Atkins diet protocol

- Copy of a carbohydrate counting guide (paperback) provided.
- Carbohydrates described in detail and restricted to 10 g per day for the first month for children, 15 g per day for adults. Carbohydrates can be increased after 1–2 months in most patients.
- Fats (e.g., 36% heavy whipping cream, oils, butter, mayonnaise) encouraged.
- Clear, carbohydrate-free fluids not restricted.
- Daily low-carbohydrate multivitamin and calcium supplementation.
- Urine ketones checked weekly for the first 2 months and weight checked weekly throughout dietary therapy.
- Medications left unchanged for at least the initial month, but reformulated if necessary to tablet or sprinkle (nonliquid) preparations.
- Complete blood count, liver and kidney functions, urine calcium and urine creatinine, and fasting lipid profile at baseline, 3, and 6 months.
- Discontinue the diet if ineffective after 2–3 months.

In the first prospective study of this diet, 20 children with intractable seizures were started on a 10 g per day protocol, which is described in Table 3.2.[9] Two-thirds of children demonstrated >50% reduction in seizures, with half (7 of 20) having >90% reduction.[9] Although large urinary ketosis occurred rapidly in all children, it decreased over time, yet surprisingly did not correlate with a loss of efficacy. Nine children were able to successfully reduce their anticonvulsants. The diet was well-tolerated, and the majority of children gained weight. Blood urea nitrogen increased significantly and total cholesterol trended upward from 192 to 221 mg/dL (although this was not statistically significant). A second randomized, crossover study of 20 children published in 2007 continued to show benefits from this diet.[14] This study demonstrated that 10 g of carbohydrate per day was the most effective starting carbohydrate limit, but could be increased to 20 g per day after 3 months without resultant loss of seizure control if achieved.[14]

Lastly, a study of the modified Atkins diet for 30 adults with intractable epilepsy demonstrated 47% of adults aged 18–53 years had at least a 50% reduction in seizures after 1–3 months, with 30% benefiting after 6 months.[13] Weight loss was 6.8 kg over a 3- to 6-month period, and was a welcome "side effect" for many overweight adults. The diet was more restrictive than for children, with slightly less than half of the patients completing the 6-month study. However, all adults with a significant response to the diet improved by 2 months, typically within 2 weeks. Considering the restrictiveness of this approach, we now recommend that adults discontinue the diet if not successful after this short time period (2 months).

LOW-GLYCEMIC INDEX TREATMENT

There is also recent evidence that a low-glycemic index treatment, similar in many ways to the South Beach diet, can be helpful for seizure control as well.[17] This diet is perhaps even less restrictive than the modified Atkins diet and does not induce similar levels of ketosis, possibly acting by stabilizing serum glucose. Foods are still relatively high in fat and protein, but allow 40–60 g of low-glycemic (glycemic index <50) carbohydrates, and calories are only roughly monitored as well. In a study of 20 patients aged 5–34 years, 50% had a >90% reduction in seizures, and 25% had a 50–90% improvement.[17] Further studies of this diet are also under way.

CONCLUSIONS

Dietary therapies are a useful treatment option for both children and adults with intractable epilepsy. While diets can improve seizure control in many patients with epilepsy, certain particular epilepsy syndromes may respond better to the diet than others. Although often seen as a more "natural" treatment, side effects from the diets do occur and the complexity of the diets makes them difficult for some families. The recent emergence of "alternative" ketogenic diets such as the modified Atkins and low-glycemic index diets have also led to additional options for patients, especially adolescents and adults. Understanding the many advantages of dietary treatments for epilepsy is very important in the care of children and adults with refractory seizures, even in this era of plentiful new and old anticonvulsants.

REFERENCES

1. Bergqvist AG, Schall JI, Gallagher PR, Cnaan A, Stallings VA. (2005). Fasting versus gradual initiation of the ketogenic diet: a prospective, randomized clinical trial of efficacy. *Epilepsia* 46, 1810–1819.
2. Freeman JM, Kossoff EH, Freeman JB, Kelly MT. (2006). *The Ketogenic Diet: A Treatment for Epilepsy in Children and Others.* 4th ed., New York: Demos Medical Publishing.
3. Freeman JM, Kossoff EH, Hartman AL. (2007). The ketogenic diet: one decade later. *Pediatrics* 119, 535–543.
4. Groesbeck DK, Bluml RM, Kossoff EH. (2006). Long-term use of the ketogenic diet. *Dev. Med. Child Neur.* 48, 978–981.
5. Henderson CB, Filloux FM, Alder SC, Lyon JL, Caplin DA. (2006). Efficacy of the ketogenic diet as a treatment option for intractable epilepsy: meta-analysis. *J. Child Neur.* 21, 193–198.
6. Kang HC, Lee YM, Kim HD, Lee JS, Slama A. (2007). Safe and effective use of the ketogenic diet in children with epilepsy and mitochondrial respiratory chain complex defects. *Epilepsia* 48, 82–88.
7. Kossoff EH, Krauss GL, McGrogan JR, Freeman JM. (2003). Efficacy of the Atkins diet as therapy for intractable epilepsy. *Neurology* 61, 1789–1791.
8. Kossoff EH, Laux LC, Blackford R, Morrison PF, Pyzik PL, Turner Z, Nordli DL, Jr. (2008). When do seizures improve with the ketogenic diet? *Epilepsia* 49, 329–333.
9. Kossoff EH, McGrogan JR, Bluml RM, Pillas DJ, Rubenstein JE, Vining EP. (2006). A modified Atkins diet is effective for the treatment of intractable pediatric epilepsy. *Epilepsia* 47, 421–424.
10. Kossoff EH, McGrogan JR, Freeman JM. (2004). Benefits of an all-liquid ketogenic diet. *Epilepsia* 45, 1163.
11. Kossoff EH, Pyzik PL, McGrogan JR, Vining EP, Freeman JM. (2002). Efficacy of the ketogenic diet for infantile spasms. *Pediatrics* 109, 780–783.

12. Kossoff EH, Pyzik PL, Rubenstein JE, Bergqvist AG, Buchhalter JR, Donner EJ, Nordli DR, Wheless JW. (2007). Combined ketogenic diet and vagus nerve stimulation: rational polytherapy? *Epilepsia* 48, 77–81.

13. Kossoff EH, Rowley H, Sinha SR, Vining EP. (2008). A prospective study of the modified Atkins diet for intractable epilepsy in adults. *Epilepsia* 49, 316–319.

14. Kossoff EH, Turner Z, Bluml RM, Pyzik PL, Vining EP. (2007). A randomized, crossover comparison of daily carbohydrate limits using the modified Atkins diet. *Epilepsy and Behavior* 10, 432–436.

15. Kwiterovich PO, Vining EP, Pyzik P, Skolasky R, Freeman JM. (2003). Effect of a high-fat ketogenic diet on plasma levels of lipids, lipoproteins, and apolipoproteins in children. *JAMA* 290, 912–920.

16. Mady MA, Kossoff EH, McGregor AL, Wheless JW, Pyzik PL, Freeman JM. (2003). The ketogenic diet: adolescents can do it, too. *Epilepsia* 44, 847–851.

17. Pfeifer HH, Thiele EA. (2005). Low-glycemic-index treatment: a liberalized ketogenic diet for treatment of intractable epilepsy. *Neurology* 65, 1810–1812.

18. Sampath A, Kossoff EH, Furth SL, Pyzik PL, Vining EP. (2007). Kidney stones and the ketogenic diet: risk factors and prevention. *J. Child Neur.* 22, 375–378.

19. Stafstrom CE, and Spencer S. (2000). The ketogenic diet: a therapy in search of an explanation. *Neurology* 54, 282–283.

20. Stainman RS, Turner Z, Rubenstein JE, Kossoff EH. (2007). Decreased relative efficacy of the ketogenic diet for children with surgically approachable epilepsy. *Seizure* 16, 615–619.

21. Wheless JW. The ketogenic diet: An effective medical therapy with side effects. (2001). *J. Child Neur.* 16, 633–635.

4 Vagus Nerve Stimulation Therapy

James W. Wheless, M.D.

CONTENTS

THE VAGUS NERVE STIMULATION THERAPY SYSTEM

Vagus nerve stimulation (VNS), which attenuates seizure frequency, severity, and duration by chronic intermittent stimulation of the vagus nerve, is intended for use as an adjunctive treatment with antiepileptic drug (AED) therapies. As of January 2008, more than 45,000 patients with epilepsy have been implanted with the VNS therapy system worldwide, with approximately 30% of those patients being younger than age 18 at the time of their first implant. Approximately one-third of patients receiving VNS therapy experience at least a 50% reduction in seizure frequency with no adverse cognitive or systemic effects.[6] Moreover, clinical findings indicate that the effectiveness of VNS therapy continues to improve over time, independent of changes in AEDs or stimulation parameters.[18] Tolerance does not appear to be a factor with VNS therapy, even after extended periods of time.[26] Response to VNS therapy may be delayed for some patients. The long-term safety and effectiveness seen with this treatment have made VNS therapy a mainstream treatment option for a broad range of epilepsy patients, including children and adolescents.

The VNS therapy system consists of the implantable pulse generator and bipolar VNS therapy lead, a programming wand with software, a tunneling tool, and

FIGURE 4.1 VNS therapy generators: model 102 (right side of picture) and 103 (left side of picture).

a handheld magnet (Figure 4.1). The average battery life for the generator is approximately 7 to 10 years with normal use (but depends on stimulation parameters—that is, frequency and intensity—as well as model type).

The magnet provided to patients as part of the VNS therapy system allows on-demand stimulation, which has the potential to abort seizures, either consistently or occasionally, among some patients or caregivers who are able to anticipate the onset of their seizures. The additional stimulus train that results when the magnet is held over the generator is typically stronger than the programmed stimulus parameters. This added ability of on-demand stimulation provides a greater sense of control for patients and their caregivers over their disorder, which can help improve how they perceive their quality of life. The magnet also allows temporary interruption of stimulation if needed, particularly when singing or playing woodwind instruments or during speaking engagements. However, stopping the stimulus should be done sparingly and with care, as doing so creates the potential risk of seizures.

IMPLANTATION PROCEDURE

The implant surgery is most often performed as a day surgery under general anesthesia and typically lasts about 1 hour.[7] The pacemaker-like generator device is generally implanted in the subcutaneous tissues of the upper left pectoral region, with a lead then run from the generator device to the left vagus nerve in the neck (Figure 4.2). Two incisions are made during the procedure—one in the chest to create the generator pocket, and the other along a fold in the neck to expose the vagus nerve for placement of the electrode (Figure 4.3). The device is often turned on in the operating room or in the office immediately after surgery, generally with a low initial setting of 0.25 mA. The programming wand (Figure 4.4) is used at follow-up visits to check and fine-tune the stimulation settings according to patient comfort and level of seizure control.

Once a generator reaches end of service, another surgery is required to replace the generator. Often, an increase in seizure frequency or intensity suggests clinical end of service. The entire generator is replaced, rather than just the battery, so as to avoid opening the hermetically sealed titanium case of the generator, which could lead to a rejection reaction. The generator-replacement surgery typically lasts approximately 10 to 15 minutes and is performed as a day surgery. Because the leads remain untouched during a generator replacement, only one incision is needed. Generator replacement is recommended before the battery is completely depleted so as to prevent an interruption in treatment.

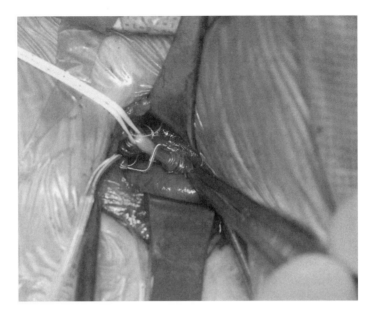

FIGURE 4.2   Lead wire starting to be placed on the left vagus nerve.

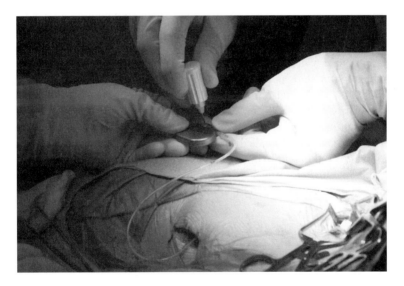

FIGURE 4.3   Implantation of model 103 for VNS therapy.

Potential Complications

One possible risk resulting from the implantation surgery is an infection at the implant site. This risk may be increased in the pediatric population because young children or patients with neurocognitive disorders may tamper with the wound before the

FIGURE 4.4 A programming wand is held by the patient over the device while a physician checks and/or adjusts stimulation parameters using a handheld computer.

incision has had time to heal properly.[8] Such infections can be treated with antibiotics, but typically lead to explantation of the device if antibiotic treatment is not effective .

The routine lead test performed during surgery also has resulted in reports of bradycardia and asystole in a small number of patients (~0.1%).[2] Neither of these cardiac events, however, has occurred after surgery during day-to-day treatment with VNS therapy, or in children; they are usually transient and self-limiting, and are rarely of clinical significance. Vocal cord paresis, although rare, can be caused by manipulation of the vagus nerve during the implant procedure, but such paresis is most often transient.

STIMULATION PARAMETERS

VNS therapy "dosing" is defined by five interrelated stimulation parameters (Figure 4.5)—output current (measured in mA), signal frequency (Hz), pulse width (μs), signal "on" time (s), and signal "off" time (s/min). The output current, signal frequency, and pulse width define how much energy is delivered to the patient, with the combination of settings for these three parameters being analogous to the size or dose of a pill. The signal "on" and "off" times constitute the duty cycle (i.e., how often the energy is delivered) and are analogous to the dosing schedule for drug therapy. An optimal dose-response relationship for VNS therapy, however, is elusive, owing in part to the intraindividual variability between patients and to the number of parameters involved in regulating the dose.

Standard parameter settings, as determined from clinical trials,[12] range from 20 to 30 Hz at a pulse width of 250 to 500 μs and an output current of 0.25 to 3.5 mA

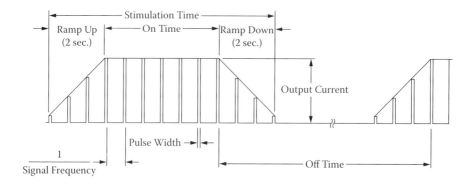

FIGURE 4.5   Stimulation parameters (all duty cycles except low output [≤10 Hz]).

TABLE 4.1
Stimulation parameter setting ranges

Parameter	Typical range	Median settings			
		Pediatric (n = 743)		Adult (n = 1486)	
		3 Months	12 Months	3 Months	12 Months
Output current	0.25–3.5 mA	1.25 mA	1.75 mA	1.00 mA	1.50 mA
Signal frequency	20–30 Hz	30 Hz	30 Hz	30 Hz	30 Hz
Pulse width	250–500 µs	500 µs	500 µs	500 µs	250 µs
Signal on time	7–270 s	30 s	30 s	30 s	30 s
Signal off time	12 s–180 min	5 min	3 min	5 min	3 min

Note: No standard settings have been defined on the basis of patient age or seizure type. The median settings shown here are taken from the VNS therapy patient outcome registry (Cyberonics, Inc., Houston, Texas).

for 30 s "on" time and 5 min "off" time (Table 4.1). Initial stimulation is set at the low end of these ranges and slowly adjusted over time and within the safety limits on the basis of patient tolerance and response. Patients should be closely monitored during the dose-adjustment phase of VNS therapy, typically every 2 to 4 weeks for the first 2 to 8 weeks following generator implantation. Once a patient responds to a tolerated dose, further parameter adjustments are performed only as clinically necessary. However, routine assessment of lead-wire integrity and generator function should be performed.

Response to VNS therapy has been shown to be age dependent, and therefore, VNS stimulus parameters may need to be adjusted differently for the pediatric patient. Several studies indicate that pediatric patients may require higher output currents (Table 4.1) than those used in adult patients to reach a therapeutic dose (2.0 to 2.5 mA compared with 1.0 to 1.75 mA, respectively), particularly when lower (≤250 µs) pulse durations are used. However, other reports indicate that clinically significant responses may occur with low stimulation intensities (1.25 to 1.50 mA).

MECHANISMS OF ACTION

The mechanisms of action of VNS therapy are not fully understood, but they are believed to be manifold, owing to the diffuse distribution of vagal afferents throughout the central nervous system, and are distinct from those of traditional AED therapy.[14] Studies suggest that altered vagal afferent activities resulting from VNS are responsible for mediating seizures.[15] Rat studies indicate that VNS activation of the locus coeruleus may be a significant factor for the attenuation of seizures. Human imaging studies also implicate the thalamus in having an important role in regulating seizure activity. The exact antiseizure role of the thalamus is likely complex, however, owing to the diffuse connections of the thalamus throughout the brain.

Imaging findings, coupled with the clinical findings that the effectiveness of VNS therapy continues to improve over time, seem to indicate that rapidly occurring subcortical effects, rather than rapidly occurring cortical effects, may be more important in the VNS antiseizure mechanism. It is believed that rapidly altered intrathalamic synaptic activities as well as other mechanisms likely occurring independently of thalamic activation comprise the therapeutic mechanisms of VNS.[28]

SEIZURE EFFICACY: CLINICAL TRIALS

Results from two randomized, placebo-controlled, double-blind trials (E03 and E05) were pivotal in demonstrating the antiseizure effect of VNS therapy.[1,11,18]

Although the controlled clinical trials did not focus specifically on the pediatric patient, the children and adolescents included in one of the five clinical studies (E04) responded at least as favorably as the adults. Of the 60 pediatric patients included in the E04 open, prospective study, 16 were younger than age 12 (mean age, 13.5 years).[13] At 3 months, the median reduction in seizure frequency was 23% ($n = 60$); for the 46 patients with follow-up data available at 18 months, the median reduction was 42%. The results, although in a much smaller group, were similar for the patients aged 11 years and younger, indicating that age does not seem to be a factor in the effectiveness of VNS therapy to control seizures.

The largest study to date to evaluate the effectiveness, tolerability, and safety of VNS therapy among pediatric patients was a six-center, retrospective study of 125 patients aged 18 years or younger (41 patients aged less than 12 years). This study showed greater reductions in seizure frequency than those found in the pediatric subgroup of the E04 clinical trial, with a median reduction in seizure frequency at 3 months of 51.5% (range, −100% to +312%; $n = 95$) and 51.0% at 6 months (range, −99.9% to +100.0%; $n = 56$). These reductions did not differ between patients with different seizure types.

SPECIAL PATIENT POPULATIONS

Although few prospective or controlled trials have been performed among pediatric epilepsy patients, the number of young patients receiving VNS therapy across the United States and Europe is growing. Observations of pediatric patients with age-related or specialized syndromes receiving VNS therapy indicate that this treatment is safe and effective across a broad range of seizure types and syndromes, independent of age. Table 4.2 shows the epilepsy syndromes, seizure types, and associated conditions where VNS therapy may be helpful. Additionally, VNS therapy also seems to be a palliative treatment option for patients who have failed cranial surgery.

Retrospective studies of the efficacy of VNS therapy among patients with Lennox–Gastaut syndrome (LGS) have shown some success in reducing seizure frequency without adverse side effects.[9] VNS therapy was performed on 50 patients from six centers (median age at implant was 13 years [range, 5 to 27 years]). This study showed that median reductions in seizure frequency at 1, 3, and 6 months of VNS therapy were 42, 58.2, and 57.9%, respectively ($n = 46$ [who had complete seizure data available]). Seizure reductions at 6 months by type showed an 88% decrease in drop attacks and an 81% decrease in atypical absence seizures. In

TABLE 4.2

Epilepsy syndromes, seizure types, and associated conditions where VNS therapy may be helpful

Simple partial seizures, simple partial seizures progressing to complex partial seizures or secondary generalization, and complex partial seizures with or without secondary generalization

Symptomatic generalized tonic–clonic seizures

Drop attacks in Lennox–Gastaut syndrome

Primary generalized epilepsy (JME)

Tuberous sclerosis complex with complex partial or generalized tonic–clonic seizures

Autism with symptomatic epilepsy

addition, improvements in quality of life with minimal and tolerable side effects from both the surgery and therapy were reported for this patient population. The most notable change in quality of life was an increase in alertness reported for more than half of the patients. Previous corpus callosotomy was not a contraindication for VNS therapy among this patient population, with the five patients who had undergone such surgery showing a 69% reduction in seizure frequency at 6 months.

VNS therapy may be an attractive treatment option among patients with developmental and behavioral comorbidities in addition to epilepsy because VNS therapy may reduce the frequency of seizures without the pharmacological side effects or interactions of additional drug therapy. Another potential benefit is the fact that VNS therapy is delivered automatically, meaning that compliance and caregiver reliance for treatment is minimized, which is particularly attractive for this patient population because many are unable to care for themselves. Studies of the effects of VNS therapy in this patient group show success with the therapy, not just with respect to seizure frequency and severity but also improvements in many areas of the patients' functional status, including alertness, mood, and daily task participation. Similar findings were obtained[10] in a retrospective study comparing outcomes of patients receiving VNS therapy living in residential treatment facilities (RTFs) with those not living in RTFs, with more improvements reported at 12 months than at 3 months, consistent with a cumulative effect of VNS therapy.

A retrospective, multicenter, open-label study of 10 patients (mean age of 13 years) with tuberous sclerosis complex (TSC) receiving at least 6 months of VNS therapy (with a mean of 22 months) found a high response rate to VNS therapy, with 9 out of 10 patients experiencing at least a 50% reduction in seizure frequency.[21] More notably, 5 of the 10 patients experienced a more than 90% reduction in seizure frequency.

Preliminary data also suggest that VNS therapy may be effective among patients with epilepsy and either autism or Landau–Kleffner syndrome (LKS), childhood disorders in which epilepsy is a prominent comorbid condition.[22] A small study of six pediatric patients (≤16 years) with hypothalamic hamartomas and refractory epilepsy indicates that VNS therapy may have the ability to independently improve behavior and, to a lesser extent, decrease seizure frequency or severity in this patient population.[20]

SAFETY

ADVERSE EVENTS

Adverse events reported with VNS therapy are generally transient and mild, and are often related to the duration and intensity of stimulation. Serious adverse events have not been reported with standard therapy, and no patients have died or had a higher mortality risk as a result of VNS therapy.[4] The most common adverse events reported during the clinical trials were mild hoarseness or voice alterations, coughing, and paresthesia (primarily at the implant site and decreasing over time) and were not considered clinically significant. Other side effects reported less frequently during these studies include dyspnea, pain, headache, pharyngitis, dyspepsia, nausea, vomiting, fever, infection, depression, and accidental injury. Not all of these side effects were related to VNS therapy. Outside the clinical trials, occasional reports of additional adverse events such as shortness of breath and vocal cord paresis have been reported, but did not result in discontinuation of therapy. Moreover, many of the side effects tend to diminish or disappear altogether as patients adjust to the stimulation therapy. If side effects persist or are bothersome to the patient, reductions in stimulation intensity or frequency often alleviate them, most of which occur only during active stimulation.[16]

Pediatric patients seem to have a higher tolerance for VNS therapy. Rare occurrences of increased salivation, increased hyperactivity, and swallowing difficulties have been reported in children. Overall, the side effects reported for pediatric patients are often mild and transient.

DEVICE SAFETY

Safety features are built into the VNS therapy system to protect patients from stimulation-related nerve injury. The primary safety feature is the "off" switch effect of the magnet. If a patient begins to experience continuous stimulation or uncomfortable side effects as a result of VNS therapy, the magnet can be held or taped over the generator to stop stimulation until the patient can visit the physician. A watchdog timer also is programmed into the device to monitor the number of pulses a patient receives. If a certain number of pulses is delivered without an "off" time, the device will turn itself off to prevent excess stimulation from potentially causing nerve injury.

Procedures such as diathermy and full-body magnetic resonance imaging (MRI) scans, which have the potential to heat the device leads around the vagus nerve and thereby result in either temporary or permanent tissue/nerve damage, are contraindicated among patients receiving VNS therapy. Patients requiring an MRI should have the procedure performed with a head coil, which has been done successfully in VNS therapy patients. As recommended by the FDA, any instructions for MRI imaging that may be in the labeling for the implant should be followed exactly, and information on the types and/or strengths of MRI equipment that may have been previously tested for interaction with the implanted device should be noted.[25] Leads or portions of leads are sometimes left in the body among patients who have had the pulse generator explanted. Therefore, it is important to get information regarding previously implanted devices as the remaining leads could possibly become heated and damage the surrounding tissue.

CANDIDATE SELECTION

Because the mechanisms of action are not well defined, the selection of patients for VNS therapy does not follow a clear set of guidelines. In addition, the clinical trials for VNS therapy could not distinguish any correlation between patient response and seizure type or etiology, age, sex, frequency of seizures, or frequency of interictal spikes on EEG to generate any obvious candidate selection criteria.[27] Therefore, similar to AED therapy, there are currently no markers to predict the success of VNS therapy on a case-by-case basis. Figure 4.6 shows a suggested treatment-sequence flowchart that could be helpful in determining which palliative surgical procedures to choose when patients are experiencing refractory seizures.

Patients of any age should be considered for VNS therapy if they experience seizures refractory to other therapies, including AEDs, the ketogenic diet, and epilepsy surgery. Preliminary data suggest that patients treated with VNS therapy earlier in the course of their epilepsy (i.e., when seizures fail to respond to treatment with two or three AEDs within 2 years of epilepsy onset) may have a higher response rate to treatment.[24]

Precautions should be taken with patients predisposed to cardiac dysfunction and obstructive sleep apnea (OSA) as stimulation may increase apneic events, and chronic obstructive pulmonary disease may increase the risk of dyspnea. Lowering the

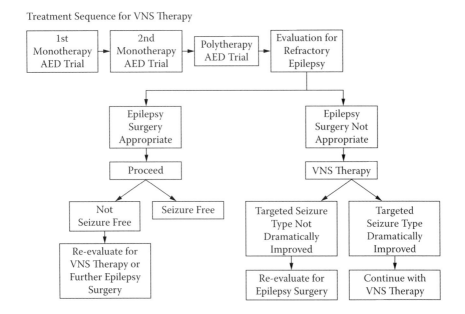

FIGURE 4.6   Suggested treatment-sequence flowchart for patients with epilepsy.

27. Wheless JW, Baumgartner J, Ghanbari C. (2001) Vagus nerve stimulation and the keto-genic diet. *Neurol. Clin.* 19: 371–407.

28. Zabara J. (1985) Time course of seizure control to brief, repetitive stimuli. *Epilepsia* 26: 518.

5 Epilepsy Surgery in Children

Tobias Loddenkemper, M.D.

CONTENTS

INTRODUCTION

Epilepsy surgery in children with medically intractable epilepsy has been well established over the last 30 years.[4,18] Advances in technology, including continuous video-electroencephalogram (EEG) monitoring, magnetic resonance imaging (MRI), magnetencephalography (MEG), single photon emission computed tomography (SPECT), positron emission tomography (PET) scan, image coregistration techniques, and refinement of surgical procedures have contributed enormously to the safety and efficacy of epilepsy surgery, resulting in successful operations in younger patients and in patients with high risk of complications.[2,18] Early aggressive management and surgical intervention may not only significantly decrease seizure frequency but also improve associated developmental delay.[4,8] Nevertheless, not all epilepsy patients are eligible for epilepsy surgery, and careful candidate selection therefore remains crucial.

CANDIDATE SELECTION FOR EPILEPSY SURGERY

A child with medically refractory epilepsy is often a potential candidate for epilepsy surgery. Intractability must be demonstrated prior to surgery by failed attempts to control seizures with antiepileptic medications. Additionally, a localizable brain abnormality should be suspected, and it is presumed that the remainder of the brain is either normal or relatively normal, compared to the area targeted for resection. Lastly, the anticipated neurological deficit after resection should be acceptable for the child and his/her family. Significant postoperative dysfunction can occur, especially when the epileptogenic zone is either in or adjacent to the eloquent cortex.

MEDICAL INTRACTABILITY

Theoretically, medical intractability arises from a combination of the severity of the epilepsy and the degree of effectiveness of the medications. The determination of an adequate response depends on several variables, but most importantly on the degree of seizure reduction (in terms of both frequency and severity) and on the side effects of the medications employed.

Patients often benefit from earlier and more aggressive treatment of their epileptic seizures. The pediatric brain has a limited window of developmental plasticity, and definitive treatment should not be substantially delayed due to prolonged trials of many available antiepileptic drugs. Therefore, it is recommended that careful selection of antiepileptic medications be made, based on age, seizure type, clinical presentation, and side-effect profile. Unfortunately, despite the availability of many new antiepileptic medications over the past 15 years, the number of medically intractable epilepsy patients has not diminished.

Kwan and Brodie[6] demonstrated that 47% of patients with newly diagnosed epilepsy were fully controlled with one antiepileptic medication, 13% had seizure control with the second medication, but only 4% with a third or multiple medications. In their cohort, 36% of patients remained medically intractable.[6] This study highlights the fact that multiple additional medication trials predict a gradually diminishing chance of controlling seizures. As a minimal condition to be considered for possible epilepsy surgery, patients must have failed valid trials with at least two or three major antiepileptic medications. However, if a preoperative evaluation reveals a clear lesion remediable by epilepsy surgery—with concordant findings based on diagnostic studies—a more rapid decision for surgery can be made. Ultimately, one must be careful in assessing medical intractability, as patients may exhibit paroxysmal nonepileptic events, may be noncompliant with medications, or may be improperly treated.

DELINEATION OF THE EPILEPTOGENIC ZONE AND ELOQUENT AREAS OF CORTEX

The preoperative workup aims at identification of the "epileptogenic zone," which is defined as the area of cortex primarily responsible for the generation of clinical seizures.[13] The goal of epilepsy surgery is the resection or isolation (through disconnection) of the epileptogenic zone and the preservation of eloquent cortex. Seizure freedom after resection is therefore regarded as ultimate proof of the successful localization of the epileptogenic zone.

The epileptogenic zone can be identified using clinical neurophysiologic tools, along with both structural and functional neuroimaging studies. This preoperative workup for epilepsy surgery includes a detailed history, EEG and continuous video-EEG monitoring, MRI, and possibly additional localizing techniques such as PET, SPECT, or MEG. When noninvasive studies fail to adequately localize the epileptogenic zone, invasive monitoring with intracranial grids and strips may be recommended. These techniques may be supplemented by additional functional tests such as neuropsychological testing, intracarotid amobarbital testing, functional MRI, intracranial stimulation, and other techniques that may assist in localization of cortical function and eloquent areas.

ESTIMATION OF THE EPILEPTOGENIC ZONE

History of seizure semiology and information on possible triggers and risk factors, such as febrile seizures, infections, trauma, tumors, stroke, family history, development, and other historical details can provide important clues toward localization and lateralization of the seizure focus. Focal examination findings can further corroborate historical information. Routine EEG and continuous video-EEG monitoring can provide additional information on clinical presentation of seizure semiology and localization of nonepileptic EEG abnormalities, interictal spikes or sharp waves (i.e., irritative zone), as well as an ictal onset zone if seizures are recorded successfully. This information can further confirm the locus of a suspected epileptogenic zone. Neuroimaging techniques, in particular MRI, may identify a lesion and may also convey clues with regard to an underlying pathology and even prognosis. In general, patients with structural lesions on neuroimaging have much better chances of becoming seizure free after epilepsy surgery.[15]

Functional neuroimaging techniques such as PET and SPECT supplement structural neuroimaging techniques, and can also help if structural neuroimaging does not reveal a focal lesion[1,11] (Figure 5.1A). Additional techniques—in particular, MEG, but also diffusion tensor-weighted imaging, spike-triggered functional MRI, and others—are gaining additional significance at selected tertiary centers.

During a second (invasive) phase of a surgical evaluation, subdural grid and depth electrode implantation or intraoperative subdural recordings can be used to test or confirm a suspected epileptogenic zone prior to resection (Figure 5.1B). Although grid recordings deliver a very good "microscopic" assessment of a certain cortical area, they should only be used during the second phase of the investigation after previously outlined techniques have supplied a "macroscopic" hypothesis, that is, a general region where the suspected epileptogenic zone may lie. Subdural grid electrodes can only provide a closer view of a certain cortical area but may miss the complete picture or may therefore deliver misleading information if the epileptogenic zone is not covered or targeted appropriately (Figures 5.2–5.4).

ESTIMATION OF CORTICAL FUNCTION AND DEVELOPMENT

Once the epileptogenic zone is approximated, overlap with eloquent cortical areas must be assessed in order to predict or prevent a possible postoperative deficit and loss of function. Tests that can assist in delineation of cortical functional areas and determine developmental function include a careful preoperative visual field assessment and clinical examination, neuropsychological assessment, intracarotid amobarbital testing for language and memory function in selected cases, and functional

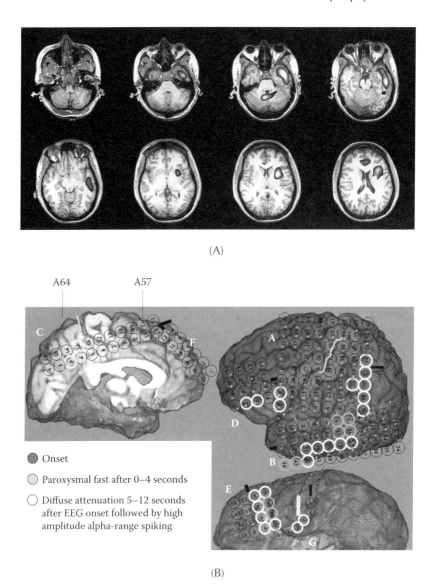

(A)

(B)

FIGURE 5.1 (A) Ictal SPECT in this 17-year-old patient with malformation of cortical development localizes seizure onset in the left temporal and insular region; (B) correlation of left posterior temporal seizure onset in the same patient on depth electrode recording (Courtesy of Dr. Andreas Alexopoulos).

MRI for language, motor, sensory, and visual function. Additional functional tests with event-related potentials and other functional imaging studies may also gain more significance in the future. Furthermore, subdural grid and strip cortical stimulation can also provide additional information with respect to cortical function.

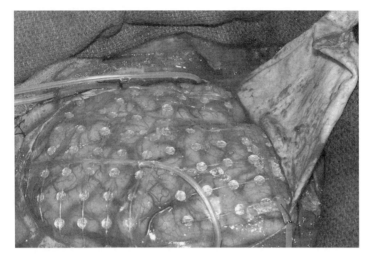

FIGURE 5.2   Intraoperative view of subdural grid electrode placement.

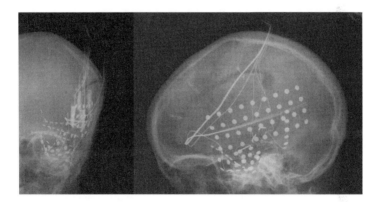

FIGURE 5.3   Anterior–posterior and lateral view skull x-rays demonstrating subdural and depth electrode placement.

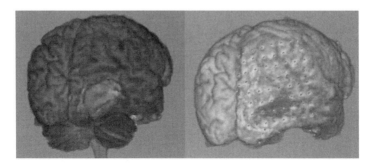

FIGURE 5.4   Coregistration of left occipital structural lesion and subdural electrode placement.

WEIGHING RISKS VERSUS BENEFITS

A preoperative workup leading to estimation of the suspected epileptogenic zone and overlap with functional areas should be discussed in a multidisciplinary patient management conference. This conference should include pediatric epileptologists and neurosurgeons, neuroradiologists, neuropsychologists and, if needed, social work and bioethics expertise. Chances of seizure freedom are then weighed against potential complications, including morbidity and mortality from surgery, risks of ongoing seizures, and potential loss of eloquent areas such as vision, motor function, memory, language, as well as neuronal plasticity. Potential developmental benefits and surgical timing should also be considered.

Studies in infants indicate that earlier epilepsy surgery may lead to improved developmental outcome.[8] In this conference, recommendations regarding surgical candidacy, further workup, type of surgery, and estimation of seizure freedom and potential deficits and complications in case of surgery are compiled. Although cost is usually not an immediate factor in the decision-making process, it is worthwhile mentioning that the costs of successful epilepsy surgery (besides benefits of decreased seizure frequency and improved quality of life), particularly in children, outweigh the long-term costs of medical care.[5]

PEDIATRIC EPILEPSY SURGERY TYPES

How often a specific type of surgical intervention occurs is highly age dependent. Whereas infants and younger children are more likely to undergo extratemporal, multilobar, or hemispheric resection (90%),[19] older children and adolescents more frequently undergo temporal resections (70%).[19]

Among others, specific surgery types include functional and anatomic hemispherectomy. Functional hemispherectomy differs from an anatomical resection in that it involves temporal lobectomy, central frontoparietal cortex and insular resection, and disconnection of frontal and parietooccipital lobes. Functional hemispherectomy decreases the risk of postoperative hydrocephalus and superficial hemosiderosis after surgery as compared to anatomic hemispherectomy.[7] Another approach is the extratemporal focal resection, which may be tailored according to extent of epileptogenic zone and eloquent cortex. Temporal resections may also be modified and can include lesionectomies, or resection of the mesial or lateral temporal lobe structures. Other techniques, such as corpus callosotomy or multiple subpial transections (a technique of selective interruption of horizontal intracortical fibers[10]), may be used in selected cases where focal resection may be difficult due to the presence of eloquent cortex.

PATHOLOGIES AND ETIOLOGIES

Frequent pathologies seen in infants and children who are candidates for epilepsy surgery include hypoxic-ischemic injury and stroke, malformation of cortical development, or dysembryoplastic neuroepithelial tumors such as ganglioglioma and gangliocytoma. Also considered should be tuberous sclerosis complex, vascular malformations, and Sturge–Weber syndrome. Mesial temporal sclerosis is rare in infants but is becoming increasingly recognized in older children and adolescents.[14] Additionally, Rasmussen's encephalitis—a focal chronic encephalitis characterized by progressive hemiparesis, hemianopia, and pharmacologically intractable seizures, and at times presenting as focal clonic status epilepticus (i.e., epilepsia partialis continua) and unilateral hemispheric atrophy—can be encountered.

OUTCOME AFTER EPILEPSY SURGERY AND PREDICTORS

Recently, the first randomized controlled trial of epilepsy surgery in adults demonstrated superiority of temporal lobe epilepsy surgery over medical management.[17] No similar randomized controlled trials involving children have been published. Nevertheless, seizure frequency rates following pediatric epilepsy surgery are encouraging. Seizure freedom in major epilepsy surgery series in infants ranges from 60–65%,[1,2,19] in children from 59–67%,[12,19] and in adolescents reaches up to 69%.[19] This mirrors the seizure freedom rate of 64% in a major adult epilepsy surgery series.[3] Overall, patients with tumors or hippocampal sclerosis tend to do better than patients with malformation of cortical development. Mortality from pediatric epilepsy surgery was 1.3% in well-established centers.[12,19]

Completeness of resection of the epileptogenic lesion (i.e., resection of a structural lesion and an area of interictal and ictal intracranial EEG findings) predicted good outcome in patients under 12 years.[12] In a review of adult and pediatric epilepsy surgery series, seizure freedom was predicted by a history of febrile seizures, mesial temporal sclerosis, abnormal MRI findings, concordance of EEG and MRI, and larger extension of resection. Postoperative seizures were more likely in patients undergoing intracranial recordings and in patients that had postoperative interictal epileptiform discharges.[16] Early postoperative seizures within 24 hours of epilepsy surgery also predicted a higher rate of recurrence of epilepsy.[9]

CONCLUSIONS

Despite the development of new antiepileptic medications, a considerable number of pediatric epilepsy patients remain pharmacologically intractable, posing significant risks to development and overall well being. Many of these medically intractable patients are eligible for epilepsy surgery, and surgery can frequently lead to seizure freedom or can at least reduce seizure frequency and severity. Careful evaluation and assessment of the epileptogenic area and eloquent cortex is warranted to ensure optimal outcomes and to prevent postoperative deficits. Epilepsy surgery in childhood takes advantage of neuronal plasticity, resulting in a greater chance of restoration of cortical functions, as well as potential recovery from developmental delays during a critical period of development. Advances in diagnostic tools to delineate the epileptogenic zone and eloquent areas may further improve outcomes following epilepsy surgery. Additional surgical techniques, such as neurostimulation, may also gain importance in selected cases. Epilepsy surgery is not an ideal solution to correct the underlying neuronal mechanisms responsible for medically refractory epilepsy, but it will remain the standard of care in selected patients until more definitive correction of the underlying causes—possibly through neurogenetic engineering techniques, more selective neurostimulation, or focal application of remedies via microcatheters and other techniques—become available.

REFERENCES

1. Chugani HT, Shewmon DA, Shields WD. et al. (1993) Surgery for intractable infantile spasms: neuroimaging perspectives. *Epilepsia* 34(4): 764–771.
2. Duchowny M, Jayakar P, Resnick T. et al. (1998) Epilepsy surgery in the first three years of life. *Epilepsia* 39(7): 737–743.
3. Engel J, Jr. (1996) Surgery for seizures. *N. Engl. J. Med.* 334(10): 647–652.
4. Falconer MA. (1970) Significance of surgery for temporal lobe epilepsy in childhood and adolescence. *J. Neurosurg.* 33(3): 233–252.
5. Keene D, Ventureyra EC. (1999) Epilepsy surgery for 5- to 18-year old patients with medically refractory epilepsy—is it cost efficient? *Childs Nerv. Syst.* 15(1): 52–54.
6. Kwan P, Brodie MJ. (2000) Early identification of refractory epilepsy. *N. Engl. J. Med.* 342(5): 314–319.
7. Lee JY, Adelson PD. (2004) Neurosurgical management of pediatric epilepsy. *Pediatr Clin. North Am.* 51(2): 441–456.
8. Loddenkemper T, Holland KD, Stanford LD, Kotagal P, Bingaman W, Wyllie E (2007) Developmental outcome after epilepsy surgery in infancy. *Pediatrics* 119(5): 930–935.
9. Mani J, Gupta A, Mascha E et al. (2006) Postoperative seizures after extratemporal resections and hemispherectomy in pediatric epilepsy. *Neurology* 66(7): 1038–1043.
10. Morrell F, Whisler WW, Bleck TP. (1989) Multiple subpial transection: a new approach to the surgical treatment of focal epilepsy. *J. Neurosurg.* 70(2): 231–239.
11. Otsubo H, Hwang PA, Gilday DL, Hoffman HJ. (1995) Location of epileptic foci on interictal and immediate postictal single photon emission tomography in children with localization-related epilepsy. *J. Child. Neurol.* 10(5): 375–381.
12. Paolicchi JM, Jayakar P, Dean P et al. (2000) Predictors of outcome in pediatric epilepsy surgery. *Neurology* 54(3): 642–647.
13. Rosenow F, Luders H. (2001) Presurgical evaluation of epilepsy. *Brain* 124(Pt 9): 1683–1700.
14. Smith ML, Elliott IM, Lach L. (2004) Cognitive, psychosocial, and family function one year after pediatric epilepsy surgery. *Epilepsia* 45(6): 650–660.
15. Spencer SS, Berg AT, Vickrey BG. et al. (2003) Initial outcomes in the multicenter study of epilepsy surgery. *Neurology* 61(12): 1680–1685.
16. Tonini C, Beghi E, Berg AT. et al. (2004) Predictors of epilepsy surgery outcome: a meta-analysis. *Epilepsy Res* 62(1): 75–87.
17. Wiebe S, Blume WT, Girvin JP, Eliasziw M. (2001) A randomized, controlled trial of surgery for temporal-lobe epilepsy. *N. Engl. J. Med.* 345(5): 311–318.
18. Wyllie E, Bingaman WE. (2001) Epilepsy Surgery in Infants and Children. In Wyllie E, editor. *The Treatment of Epilepsy: Principles and Practice*. Philadelphia: Lippincott, Williams & Wilkins, 1161–1173.
19. Wyllie E, Comair YG, Kotagal P, Bulacio J, Bingaman W, Ruggieri P. (1998) Seizure outcome after epilepsy surgery in children and adolescents. *Ann. Neurol.* 44(5): 740–748.

6 Status Epilepticus

James J. Riviello, Jr., M.D.

CONTENTS

Status epilepticus (SE) is defined as a seizure lasting 30 minutes or more of either continuous seizure activity or two or more sequential seizures with persistent altered awareness in between.[9] It is a life-threatening medical emergency requiring prompt recognition and treatment, starting with the basic principles of neuroresuscitation—the A, B, Cs—followed by a planned treatment protocol. SE is not a specific disease itself and may occur during the course of epilepsy or secondary to a central nervous system (CNS) insult. Proper management requires the identification and treatment of the precipitating cause in order to facilitate seizure control and prevent ongoing neurologic injury.

Our SE cases will review the diagnosis and treatment of both convulsive SE (CSE) and nonconvulsive SE (NCSE), including the neurodiagnostic testing needed and the treatment of refractory SE. The first case will cover the differential diagnosis, diagnostic approach, and treatment of SE in general; subsequent chapters will review specific situations.

CASE PRESENTATION

A boy had motor delay and a left hemiparesis. An initial head computed tomography (CT) scan at one year of age demonstrated an atrophic right hemisphere. He did well until 7 years of age when he developed a focal motor seizure of the left side that lasted 10 minutes. Brain MRI showed a large right parietal porencephalic cyst. He was started on oxcarbazepine, but the seizures recurred. The stereotyped events all consisted of an aura, with nausea, which could progress to altered awareness with automatisms. His recurrent seizures were initially short, and several antiepileptic drugs were subsequently tried. His current regimen was oxcarbazepine with topiramate. At nine years of age, during an episode of influenza with associated vomiting, he developed several seizures in one day, again with focal motor movements on the left side, associated with altered awareness. Antiepileptic drug levels were checked; the oxcarbazepine level was low, resulting in an increase in the dose. During the night, his parents heard him making a gurgling noise and found him experiencing a generalized tonic–clonic seizure. This seizure did not stop, and emergency medical service (EMS) was called. He was transported to a local emergency department where he received a dose of lorazepam, but the seizure persisted. Initial evaluation showed hyponatremia. He was intubated to protect his airway, received a second dose of lorazepam with a cessation of convulsive movements, but remained post-ictal. A repeat head CT scan was done, followed by a lumbar puncture.

DIFFERENTIAL DIAGNOSIS/DIAGNOSTIC APPROACH

In this specific case, SE occurred during an intercurrent influenza illness. This episode of SE most likely represents an exacerbation of the underlying seizure disorder during an intercurrent illness. This is a common occurrence, although the differential diagnosis must include influenza encephalitis or meningitis. For epilepsy in general, and especially for SE, it is critical to identify a precipitating cause, even in a patient with an underlying seizure disorder, because a specific precipitant may require specific treatment.

Blood chemistries should be performed in patients at risk for metabolic abnormalities, such as those with gastroenteritis. Although the overall yield is only 6%, it should be considered in at-risk patients. Hyponatremia itself may precipitate seizures and may be caused by carbamazepine and oxcarbazepine, whereas acidosis may occur with topiramate. Neonates often present with hypoglycemic and hypocalcemic seizures, so these should be evaluated for in this age group. A lumbar puncture is diagnostic for a CNS infection in 12% of patients with SE and should be performed in patients with concerns for intracranial infection, such as those with nuchal rigidity, fever, or unexplained encephalopathy. Other abnormal diagnostic studies included low antiepileptic drug levels (32%), ingestion of drugs or toxins (3.6%), and inborn error of metabolism (4.2%). An EEG detects epileptiform abnormalities in 43% of patients, and should be performed in all patients who remain unresponsive to evaluate for nonclinical status epilepticus.[12]

In the North London Status Epilepticus Surveillance Study (NLSTEPSS), the first prospective study of only children with SE, 33% of new onset SE cases were prolonged febrile seizures, and another 16% had an acute CNS insult.[1] In the Richmond study, a medication change occurred in 20%, followed by a metabolic abnormality in 8%, anoxia and CNS infection in 5%, trauma and vascular etiologies in 3.5%, and intoxications in 2.5%.[2]

The most common classification system for SE is based on the etiology: symptomatic, remote symptomatic, remote symptomatic with an acute precipitant, and febrile SE.[1,12] In this case, an acute precipitant is likely. Using a retrospective analysis of SE, the practice parameter from the American Academy of Neurology (AAN) on the diagnostic assessment of the child with SE identified an acute precipitant in only 1%.[12] However, the NLSTEPSS identified an acute on remote symptomatic cause 16%.[1] It is therefore important to identify a precipitant and treat appropriately because controlling seizure cessation with antiepileptics does not treat the precipitating cause.

Finally, for SE diagnosis, when is neuroimaging needed? Neuroimaging is needed with new-onset SE, or when the baseline neurologic examination has changed, such as with focality, especially if new.[7] In the AAN practice parameter, neuroimaging abnormalities were detected in 8% of the children.[12] These abnormal findings may be related to the cause of the underlying epilepsy but may not have precipitated the acute episode of SE. In new-onset SE or with a new focally abnormal neurologic examination, if there is concern for a CNS infection, neuroimaging should be done prior to a lumbar puncture.

TREATMENT

The strict criterion for SE is 30 minutes of either CSE or serial seizures without recovery of consciousness in between the seizures. However, we do not wait 30 minutes to treat SE because there is concern about the potential for brain injury. A recent "operational definition" of SE recommended treatment after 5 minutes for either (1) 5 minutes or more of a continuous seizure, or (2) two or more discrete seizures with incomplete recovery of consciousness in between.[9]

What are the appropriate antiepileptic medications for status epilepticus, and is there a specific sequence in which these should be given? Standard treatment guidelines are needed, but in a recent UK survey, only 12% had a planned protocol.[13] Evidence-based guidelines are important to standardize care, analyze outcomes, and improve treatment. These are extremely important for providing treatment consistency.

We have included the current clinical practice guideline for the treatment of SE at Texas Children's Hospital[14] (see Table 6.1). Especially in a younger child, consider pyridoxine, 100 to 200 mg, if SE does not respond to protocol.

What is the response to treatment of SE? In one series, 85% responded to the first dose of a benzodiazepine; although guidelines typically use a repeat benzodiazepine, this only controlled an additional two cases.[5] In another series, 73% responded to either IV or rectal diazepam, and 16.5% responded then to either phenobarbital or phenytoin.[3,4]

Could treatment be done at home to prevent this? In a case with recurrent seizures, are there medications that can be given at home, and when should these be considered? Several medications can now be given at home. Intravenous diazepam solution can be given rectally. The most commonly used is a specific diazepam rectal gel, Diastat®(Valeant Pharmaceuticals International), which has the advantage that it is premixed and much easier for families to administer. Intranasal midazolam has also been used, whereas lorazepam is available in a solution, and clonazepam is available in a wafer. In a recent study of Diastat, seizures were controlled in 84%, avoiding a visit to the emergency department. At-home treatments also reduce parental anxiety.[11]

TABLE 6.1

Practice guideline for the treatment of SE at Texas Children's Hospital

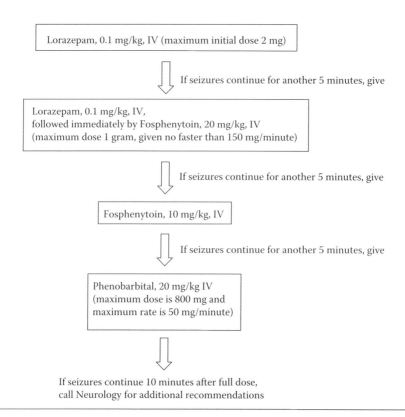

Lorazepam, 0.1 mg/kg, IV (maximum initial dose 2 mg)

If seizures continue for another 5 minutes, give

Lorazepam, 0.1 mg/kg, IV,
followed immediately by Fosphenytoin, 20 mg/kg, IV
(maximum dose 1 gram, given no faster than 150 mg/minute)

If seizures continue for another 5 minutes, give

Fosphenytoin, 10 mg/kg, IV

If seizures continue for another 5 minutes, give

Phenobarbital, 20 mg/kg IV
(maximum dose is 800 mg and
maximum rate is 50 mg/minute)

If seizures continue 10 minutes after full dose,
call Neurology for additional recommendations

Why is it important to control SE as soon as possible? Seizures, especially SE, increase cerebral metabolic demands, especially for glucose and oxygen. Initially, compensatory mechanisms are able to meet these demands; however, as the duration of SE increases, hypoxemia, hypercarbia, hypotension, and hyperthermia occur. A mismatch develops between the ongoing metabolic needs of the brain and a possible decrease in cerebral blood flow and brain depletion of glucose and oxygen.[8] Also, as SE duration increases, alterations in receptor function may decrease the efficacy of antiepileptics.[6]

OUTCOME

The outcome and mortality of SE are related to the etiology. In seizures that last greater than 5 minutes, 50% of febrile seizures resolved spontaneously, whereas no acute symptomatic seizures did.[4] Mortality rates vary from 3 to 11%.[12] In the Richmond study, the overall mortality was 6%, but when age-stratified, mortality in the first year was 17.8%, and 24% in the first six months. The increased mortality within the first year is due to a higher incidence of acute symptomatic SE in the youngest children.[10]

REFERENCES

1. Chin RFM, Neville BGR, Peckham C, Bedford H, Wade A, Scott RC. (2006) Incidence, cause, and short-term outcome of convulsive status epilepticus in children: prospective, population-based study. *Lancet* 368: 222–229.
2. DeLorenzo RJ, Towne AR, Pellock JM, Ko D. (1992) Status epilepticus in children, adults, and the elderly. *Epilepsia* 33(Suppl. 4): 15–25.
3. Eriksson K, Koivikko M. (1997) Status epilepticus in children: aetiology, treatment and outcome. *Dev. Med. Child Neurol.* 39:652–658.
4. Eriksson K, Metsaranta P, Huhtala H, Auvinen A, Kuusela A-L, Koivikko M. (2005) Treatment delay and the risk of prolonged status epilepticus. *Neurology* 65: 1316–1318.
5. Garr RE, Appleton RE, Robson WJ, Molyneux EM. (1999) Children presenting with convulsions (including status epilepticus) to a paediatric accident and emergency department: an audit of a treatment protocol. *Dev. Med. Child Neurol.* 41: 44–47.
6. Goodkin HP, Joshi S, Kozhemyakin M, Kapur J. (2007) Impact of receptor changes on treatment of status epilepticus. *Epilepsia* 48(Suppl. 8): 14–15.
7. Harden CL, Huff JS, Schwartz TH et al. (2007) Therapeutics and Technology Assessment Subcommittee of the American Academy of Neurology. Reassessment: neuroimaging in the emergency patient presenting with seizure (an evidence-based review): report of the Therapeutics and Technology Assessment subcommittee of the American Academy of Neurology. *Neurology* 69: 1772–80.
8. Lothman E. (1990) The biochemical basis and pathophysiology of status epilepticus. *Neurology* 40(Suppl. 2): 13–23.
9. Lowenstein DH, Bleck T, Macdonald RL. (1999) It's time to revise the definition of status epilepticus. *Epilepsia* 40: 120–122.
10. Morton LD, Garnett LK, Towne AR et al. (2001) Mortality of status epilepticus in the first year of life. *Epilepsia* 42(Suppl. 7): 164–165.
11. O'Dell C, Shinnar S, Ballaban-Gil KR, Hornick M, Sigalova M, Kang H, Moshe SL. (2005) Rectal diazepam gel in the home management of seizures in children. *Pediatr. Neurol.* 33: 166–172.
12. Riviello JJ, Ashwal S, Hirtz D et al. (2006) Practice Parameter: Diagnostic assessment of the child with status epilepticus (an evidence-based review): Report of the Quality Standards Subcommittee of the American Academy of Neurology and the Practice Committee of the Child Neurology Society. *Neurology* 67: 1542–1550.
13. Walker MC, Smith SJ, Shorvon SD. (1995) The intensive care treatment of status epilepticus in the UK: Results of a National survey and recommendations. *Anaesthesia* 50: 130–135.
14. Wilfong A, McPherson M. (2007) Status Epilepticus. Clinical Practice Guidelines: Texas Children's Hospital, Houston, TX.

Section 2

The Neonate

7 Benign Familial Neonatal Seizures

Eric Marsh, M.D., Ph.D. and
Edward C. Cooper, M.D., Ph.D.

CONTENTS

CASE PRESENTATION

A baby boy was born vaginally at 39 and 4/7th weeks to a G2 P2002 mother after an uncomplicated pregnancy and delivery. The child stayed in the newborn nursery for two days without incident and was discharged home with his mother. On the sixth day of life, four days after being discharged from the nursery, the mother found her baby lying in the crib eyes open and deviated up and to the right with his whole body stiff. The mother reached for the child who remained stiff for about 1 minute and developed perioral cyanosis. While in the mother's arms, the child let out a brief cry, his body relaxed, and the color returned to his face. The parents called 911 and the child was taken to the emergency room for evaluation. On arrival to the emergency room, the patient was found to be afebrile, with normal vitals and clear rhinorhea. General medical exam and neurological exam were normal. A single 1 × 1 centimeter hypopigmented lesion was found on the chest. Shortly after arrival to the emergency room, the child had a second event, identical to the first, consisting of tonic body stiffening, eyes deviated, and perioral cyanosis. A workup, including lumbar puncture, a complete blood count, and electrolytes, was normal. Urine, cerebrospinal fluid (CSF), and blood cultures along with herpes simplex virus polymerase chain reaction (HSV PCR) were sent. A head computed tomography (CT) scan was performed and was normal. The child was started on phenobarbital (20 mg/kg loading dose), ampicillin, gentamycin, and acyclovir and admitted to the neonatal intensive care unit.

Upon admission, further history was obtained. The child had been well since birth, except for the development of mild nasal congestion. The patient's mother denies any complications of the pregnancy or delivery. Family history was significant for a maternal cousin who developed seizures shortly after birth after a "difficult" delivery. This cousin did well after discharge from the hospital, has had no subsequent seizures, and is off medication. A vague history of a maternal great-aunt with seizures shortly after birth was given by the baby's maternal grandmother. There also was a maternal second cousin diagnosed with benign Rolandic epilepsy. No history of seizures or other neurological issues was reported on the paternal side.

After being admitted, all the child's vitals remained normal. A few hours after the loading dose in the emergency room, he had a third seizure, identical to the previous two events. An electroencephalogram (EEG) was obtained and showed a normal background with rare left-sided centrotemporal spikes. An additional 10mg/kg load of phenobarbital was given and no further seizures were documented. After 48 hours of negative cultures, normal newborn screening labs, and normal metabolic screening tests, the patient was discharged home with his parents with a diagnosis of benign familial neonatal seizures (BFNS, also called benign familial neonatal convulsions or BFNC) suspected. The patient was treated with phenobarbital for 3 months with no further seizures. As of age 2 there have been no recurrent seizures, with the child having a normal developmental course.

DIFFERENTIAL DIAGNOSIS

Neonatal seizures are usually an ominous sign of significant intracranial pathology, and BFNS are a diagnosis of exclusion. In the absence of an exceptionally clear family history of dominantly inherited benign neonatal seizures, it is imperative to perform a complete workup for infectious, hypoxic–ischemic (HIE), hemorrhagic, prior ischemic, traumatic, developmental malformations, and toxicological and metabolic causes. Among the most common causes are perinatal infections (due to herpes simplex virus, Group B streptococcus, *E. coli*, *Listeria monocytogens*), congenital infections (i.e, toxoplasmosis, rubella, cytomegalovirus, and HIV), HIE, or prior ischemia. If these potentially morbid causes of neonatal seizures can be ruled out, nonepileptic causes such as benign sleep myoclonus, benign tremulousness, and reflux-related movements should also be considered. If the EEG shows the events to be seizures, then the entity of benign idiopathic neonatal (and infantile) convulsions should also be considered. A history of neonatal convulsions in one parent and his or her family, along with a completely normal interictal neurological exam, supports a diagnosis of BFNS.

DIAGNOSTIC APPROACH

Neonatal seizures should be assessed with real urgency, with the initial workup performed while treatment is initiated, to cover the suspected causes and to prevent further seizures. As infectious causes are the most serious and are readily treatable, a CBC, electrolytes, and blood and urine cultures, and a lumbar puncture (LP) should be immediately performed. A CT scan of the head should be the first imaging study performed to rule out bleeds, infarction, hydrocephalous, some TORCH infections, and structural anomalies. If these tests are normal, a metabolic etiology should be ruled out by metabolic testing, including serum amino acids, urine organic acids, lactate, pyruvate, and CSF glucose and lactate/pyruvate. Additionally, an EEG should be performed to look for subclinical seizures, status epilepticus, nonepileptic events, and the diagnosis of BFNS.

In families with a very clear history of BFNS, a workup could be limited to ruling out infectious, hemorrhagic, HIE, and traumatic causes. A strong family history of BFNS does not rule out these other etiologies, which could coexist. The EEG in BFNS can be normal interictally, or can be slow, discontinuous, focal, multifocal, or show a "theta-point-alternant pattern." This is an unreactive, discontinuous EEG, predominantly consisting of theta activity with intermixed sharp waves and interhemispheric asynchrony. It is exhibited in benign idiopathic (i.e., nonfamilial) neonatal seizures (often called "fifth day fits"), BFNS, and sometimes in symptomatic seizures related to underlying injury.

TREATMENT AFTER DIAGNOSIS

There is no consensus on the treatment for benign familial neonatal seizures. No currently approved medications have been studied for this condition. Phenobarbital, 20 mg/kg loading dose, followed by 3–10 mg/kg/day daily dosing, would be considered first line. A level between 20 and 40 mcg/mL should be targeted. If frequent seizures occur and do not respond to the phenobarbital, benzodiazepines or phenytoin should be used. The question of whether or not BFNS should be treated has been raised. Because the diagnosis is often uncertain until the child has outgrown the syndrome and developed normally, our view is that treatment to stop recurrent seizures is warranted.

LONG-TERM OUTCOME

The outcome for infants that experience BFNS is generally excellent. In a large majority of cases, seizures stop completely within the first 2–3 months (often within the first month), and infants have a normal subsequent course. Approximately 10–16% of patients experience at least one seizure later in life, often Rolandic in location and semiology (see Chapter 20). Better genetic understanding of BFNS (discussed in the following text) has introduced a slight note of caution, however. It has become clear that, occasionally, more severely affected individuals can be found in families where other affected patients have typical BFNS. Prior to the availability of genetic testing, such families were excluded by definition from the BFNS cohort and thought to have another disorder, but now a few have been linked to mutations in the ion channel genes that cause typical BFNS. A greater number of such families will have to be followed and subjected to genetic testing before this issue can be clarified.

PATHOPHYSIOLOGICAL BASIS OF BENIGN FAMILIAL NEONATAL SEIZURES

Most cases of BFNS are due to mutations in two genes, *KCNQ2* and *KCNQ3*, which encode subunits of a type of voltage-gated potassium ion (Kv) channel. Of about 70 BFNS families so far studied genetically, 60% have mutations in *KCNQ2*, and 5% have mutations in *KCNQ3*. The cause in the remaining cases is unknown. Some are likely due to mutations in portions of the *KCNQ2* and *KCNQ3* genes that do not encode amino acids but may affect channel expression (e.g., enhancers, promoters, introns). It is possible that one or more additional BFNS genes remain undiscovered.

All Kv channel proteins form by the coassembly of four subunit polypeptides. In the brain, some channels are formed by four copies of *KCNQ2*, whereas others are formed by coassembly of *KCNQ2* and *KCNQ3*, likely in a 2:2 ratio. These channels (sometimes alternatively referred to as Kv7 channels or "M-channels") open in response to membrane depolarization. The opening allows K^+ to leave the cell due to its electrochemical gradient, driving the membrane potential towards hyperpolarization. This tends to keep the membrane potential below the action potential threshold, thereby limiting neuronal responses to excitatory synaptic inputs. *KCNQ2* and *KCNQ3* are localized to axons at the sites where high densities of voltage-gated Na^+ channels are clustered, such as axon initial segments and nodes of Ranvier. Axonal *KCNQ* channels both restrain action potential firing and dampen membrane potential oscillations that otherwise occur following the passage of action potentials along the axon. Indeed, some BFNS patients and their families have (clinically significant or asymptomatic) myokymia (i.e., involuntary muscle contractions caused by aberrant action potentials arising spontaneously within motoneurons axons).

BFNS mutations reduce channel function by various mechanisms, including by slowing opening rates, increasing the extent of membrane depolarization needed for channel opening, and preventing channels from being localized properly in the neuronal cell membrane. The reasons why these mutations usually cause seizures that begin within a few days after birth and remits weeks later are not fully known. One potential reason is the observation that neurotransmission using gamma-aminobutyric acid A ($GABA_A$) receptors is weakly inhibitory during the neonatal period due to the depolarized reversal potential for Cl^- in immature neurons, rendering the neonatal brain particularly dependent on potassium channel activity. A second factor may be myelination: axonal *KCNQ* channels may be especially important for suppressing excessive neuronal firing during certain stages of myelination either generally or in particular brain regions. These questions are currently under intense investigation.

Although BFNS is rare, *KCNQ* channels have emerged as rational targets for drugs used to treat more common forms of epilepsy. Several structurally unrelated types of *KCNQ* channel openers have already been described and are in various stages of preclinical and clinical trials. Ultimately, the most general importance of

BFNS may be in alerting us to these channels as potential targets for therapeutic control of excitability.

CLINICAL PEARLS

1. Benign familial neonatal seizures usually occur during the first week of life after an initial seizure-free period.
2. BFNS is considered a diagnosis of exclusion, and a thorough evaluation for other causes of neonatal seizures needs to be undertaken even in the presence of strong family history.
3. Treatment with phenobarbital should be undertaken to control acute seizures, and is often continued for the first 3 months.
4. Abnormalities in potassium channel function underlie the cause of BFNS, and creation of medications that correct this dysfunction may offer a new, effective treatment for epilepsy.

SUGGESTED READING

Anderson VE, Plouin P. (2005). Benign familial and nonfamilial neonatal seizures. In *Epileptic Syndromes in Infancy, Childhood, and Adolescence,* J. Roger, M. Bureaus, C. Dravet, P. Genton, CA Tassinari, and P. Wolf, Eds., 3–16. Montrouge: John Libbey Eurotext.

Cooper EC, Jan LY. (2003). M-channels: Neurological diseases, neuromodulation, and drug development. *Archives of Neurology* 60, 494–500.

Pan Z, Kao T, Horvath Z, Lemos J et al. (2006). A common ankyrin-G-based mechanism retains KCNQ and NaV channels at electrically active domains of the axon. *J. Neurosci* .26, 2599–2613.

Ronen GM, Rosales TO, Connolly M, Anderson VE, Leppert M. (1993). Seizure characteristics in chromosome 20 benign familial neonatal convulsions. *Neurology* 43, 1355–1360.

Singh NA, Westenskow P, Charlier C et al. (2003). KCNQ2 and KCNQ3 potassium channel genes in benign familial neonatal convulsions: expansion of the functional and mutation spectrum. *Brain* 126, 2726–2737.

Steinlein OK, Conrad C, Weidner B. (2007). Benign familial neonatal convulsions: always benign? *Epilepsy Res.* 73, 245–249.

8 Hypoxic-Ischemic Encephalopathy (Neonatal Seizures)

Mark S. Scher, M.D.

CONTENTS

CASE PRESENTATION

A 41-week-old female was born to a 23-year-old primigravida woman who experienced decreased fetal movements several days before delivery. Fetal distress, indicated by bradycardia and loss of variability on fetal monitoring, was noted during labor. A neurologically depressed female infant was delivered and was transferred to a level III neonatal intensive care unit (NICU). Intrauterine growth restriction was diagnosed based on a birthweight of 2400 g for 39 weeks gestational age. Apgar scores were 3 at 1 minute, 6 at 5 minutes, and 8 at 10 minutes. Minimal resuscitative efforts were needed. Within 7 hours after birth, the infant exhibited clinical seizures, characterized by multifocal clonic movements confirmed on EEG recordings to represent multifocal seizures (Figure 8.1). Both head computed tomography (CT) and brain magnetic resonance imaging (MRI) revealed diffuse cerebral edema. The placental examination documented vascular changes consistent with fetal thrombotic occlusive disease, including hyalinized and avascular villi (Figure 8.2). A coagulation profile documented methyltetrahydrofolate reductase deficiency with a high serum homocysteine level and homozygosity; mother was found to be a carrier. At 5 years of age, the child exhibited microcephaly, spastic quadriplegia, and intractable generalized and focal seizures that began at less than 1 month of age (despite treatment with three antiepileptic medications).

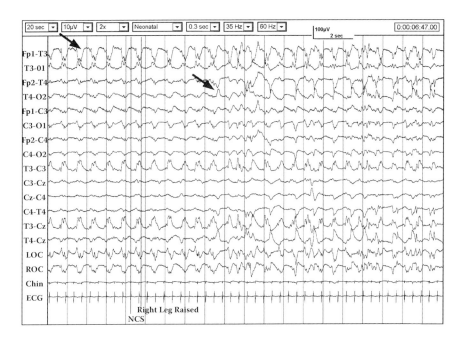

FIGURE 8.1 Tracing documenting two independently occurring electrical seizures (arrows) on a suppressed electroencephalographic background. (From Scher, M.S., Kidder, B.M., and Bangert, B.A., *J. Child Neurology*, 22(4), 396–401, 2007.)

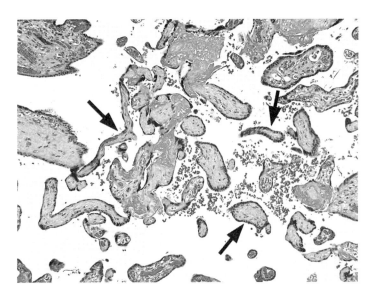

FIGURE 8.2 A placental sample consisting of a microscopic view with magnification 100X, H/E stain of placental villi showing hyalinized avascular villi (arrows) intermixed with normal villi on the maternal side of the placenta. (From Scher, M.S., Kidder, B.M., and Bangert, B.A., *J. Child Neurology*, 22(4), 396–401, 2007.)

DIFFERENTIAL DIAGNOSIS

Neonatal seizures are generally brief and subtle in clinical appearance, at times comprised of unusual behaviors that are difficult to recognize and classify. Oral–buccal–lingual or so-called "bicycling" movements are suggestive of subtle neonatal seizures, and rapid rhythmic movements of an extremity raise understandable concern for clonic seizure activity but are often a result of jitteriness that can be attenuated manually. Medical professionals vary considerably in their ability to interpret such suspicious behaviors as seizures, thus contributing to either overdiagnosis or underdiagnosis. Commonly, within neonatal intensive care units, clinical behaviors are classified as seizures without EEG confirmation. There are major pitfalls to this approach, as abnormal paroxysmal motor or autonomic behaviors may represent age and state-specific behaviors in healthy infants. Further, nonepileptic pathologic paroxysmal conditions can occur in symptomatic infants. For these reasons, confirmation of suspicious clinical events with coincident EEG recordings is strongly recommended. Although a few seizures of short durations in patients may be missed as brief random events on routine EEG studies, continuous, synchronized video/EEG/polygraphic recordings can more reliably determine beginning and endpoints for more accurate diagnoses and assessment of treatment efficacy.

Neonatal seizures are not disease-specific, and can be associated with a variety of medical conditions that occur before or during parturition, as well as during the postnatal period. Documentation of asphyxia is the most frequently diagnosed entity when seizures occur. Seizures can occur as part of an asphyxial injury associated with a neonatal encephalopathy or brain disorder. Alternatively, other etiologies for neonatal encephalopathy besides asphyxia must be considered. Seizures may also be an isolated clinical sign without other neurological signs of a postnatal encephalopathy. Although antepartum and intrapartum factors individually have low positive predictive values for predicting the occurrence of neonatal seizures, a significant cumulative risk profile can be established for variables such as antepartum maternal anemia, bleeding, and asthma, meconium-stained amniotic fluid, abnormal fetal presentation, fetal distress, and shoulder dystocia.

Postnatal medical illnesses also cause or contribute to asphyxial-induced brain injury. Persistent pulmonary hypertension of the newborn, cyanotic congenital heart disease, sepsis, meningitis, encephalitis, and primary intracranial hemorrhage are leading diagnoses. However, clinicians should consider a continuum of injury for some neonates presenting with encephalopathy beginning in the antepartum and extending into the intrapartum period. Specific clinical examination findings in the neonate suggest the occurrence of antepartum injury, even with acute signs of distress during labor, which can be superimposed on chronic antepartum injury. Intrauterine growth retardation, hydrops fetalis, or joint contractures (including arthrogryposis) are findings that support an antepartum process that later was expressed as intrapartum fetal distress followed by neonatal depression. Spasticity, often with cortical thumbing, suggests long-standing fetal neurological dysfunction, because neonates after intrapartum asphyxia are traditionally noted to be hypotonic. Intrapartum asphyxial injury can certainly add to brain injury in these children. Neonates may then exhibit signs of neonatal encephalopathy from both preexisting antepartum brain injury and subsequent intrapartum events. Neuroimaging (especially brain MRI) can define specific patterns of injury that result from asphyxia, even independent of labor and delivery, depending on when the MRI was obtained.

Placental findings can be associated with either antepartum or intrapartum conditions, as with our case history. Meconium staining through the chorionic and amnion layers suggests a more chronic asphyxial stress. Placental weights below the 10th or above the 90th percentile also suggest chronic hypoperfusion to the fetus. Examples of chronic placental injuries include vasculopathies, stromal infarction, or villus maldevelopment.

TREATMENT STRATEGY

Rapid infusion of glucose or other electrolytes should be initiated before considering antiepileptic medications. Hypoglycemia can be readily corrected by intravenous administration of 5–10 mg/kg of a 10–15% dextrose solution, followed by an infusion of 8–10 mg/kg per minute. Persistent hypoglycemia may require more hypertonic glucose solutions. Rarely, prednisone 2 mg/kg per day may be needed to establish a glucose level within the normal range. Hypocalcemia-induced seizures should be treated with an intravenous infusion of 200 mg/kg of calcium gluconate. This dosage should be repeated every 5–6 hours over the first 24 hours. Serum magnesium concentrations should also be measured because hypomagnesaemia may accompany hypocalcemia; 0.2 mg/kg magnesium sulfate should be given by intramuscular injection. Disorders of serum sodium are rare causes of neonatal seizures.

Pyridoxine dependence or deficiency is a rare, genetic condition that is eminently treatable. An injection of 50–500 mg pyridoxine is recommended during a seizure with coincident EEG monitoring to document seizure cessation. A daily dose of 50–100 mg pyridoxine should then be administered. Other biochemical deficiencies include folinic acid, biotin, or sulfite oxidase with resultant seizures. Appropriate biochemical studies are required to confirm these rare metabolic diseases.

If the decision to treat neonates with antiepileptic medications is reached, important questions must be addressed with respect to who should be treated, when to begin treatment, which drug to use, and for how long neonates should be treated. Phenobarbital and phenytoin remain the most widely used antiepileptic medications. The half-life of phenobarbital ranges from 45 to 173 hours; the initial loading dose is recommended at 20 mg/kg, with a maintenance dose of 3–4 mg/kg per day. Therapeutic levels are generally suggested to be between 16 and 40 mg/mL. There is no consensus with respect to the duration of drug maintenance. The preferred loading dose of phenytoin is 15–20 mg/kg. Serum levels of phenytoin are difficult to maintain because this drug is rapidly redistributed to body tissues. Blood levels cannot be well maintained using an oral preparation. Benzodiazepines may also be used to control neonatal seizures. The drug most widely used is diazepam, whose half-life may range from 54 hours in preterm infants to 18 hours in full-term infants. Intravenous administration is recommended because it is slowly absorbed after an intramuscular injection. Recommended intravenous doses for acute management should begin at 0.5 mg/kg. Deposition into muscle precludes its use as a maintenance antiepileptic medication as profound hypotonia and respiratory depression may result, particularly if barbiturates have also been administered.

There are conflicting reports regarding the efficacy of phenobarbital and phenytoin. Most studies apply only a clinical endpoint to seizure cessation without EEG. With EEG as an endpoint to judge cessation of seizures, neither phenobarbital nor phenytoin has been proven effective to control seizure activity in selected populations.

Drug binding in neonates with seizures has only recently been reported, and can be altered in a sick neonate with organ dysfunction. Toxic side effects may result from elevated free fractions of a drug that adversely affect cardiovascular and respiratory function. To guard against untoward effects, evaluation of treatment and efficacy must take into account both total and free antiepileptic drug fractions in the context of the newborn's progression or resolution of systemic illness. Although novel anticonvulsants with distinct mechanisms of actions have been suggested as being efficacious in uncontrolled trials or in animal studies, none has been validated or widely accepted. Clearly, there is a great need to develop more efficacious drugs to treat neonatal seizures and to prevent long-term pathophysiological consequences.

PATHOPHYSIOLOGY/NEUROBIOLOGY OF DISEASE

Hypoxia-ischemia (i.e., asphyxia) is traditionally considered the most common cause associated with neonatal seizures, but only 10% of asphyxia results from postnatal causes. Intrauterine factors prior to labor can result in fetal asphyxia without later documentation of acidosis at birth. Both antepartum and intrapartum maternal and placental illnesses associated with thrombophilia, preeclampsia, or specific uteroplacental abnormalities such as abruptio placentae or cord compression may contribute to fetal asphyxial stress and result in metabolic acidosis. Antepartum maternal trauma and chorioamnionitis are additional conditions that also contribute to the intrauterine asphyxia secondary to uteroplacental insufficiency. Intravascular placental thromboses and infarction of the placenta or umbilical cord documented after birth are markers for possible fetal asphyxia. Meconium passage into the amniotic fluid also promotes an inflammatory response within the placental membranes, potentially causing vasoconstriction and resultant asphyxia.

LONG-TERM OUTCOME

Embedded in the controversy surrounding the diagnosis of neonatal seizures is the association with poor neurologic outcome. Epilepsy occurs in 25–50% of children after neonatal seizures and is usually associated with behavioral and cognitive deficits. It is often difficult to assess the impact of the seizures themselves on neurological sequelae, independent of the underlying neurological condition or disease substrate. Further, antiepileptic drug use may induce secondary harmful effects on cardiac and respiratory function, with resultant circulatory disturbances that contribute to brain injury. Chronic antiepileptic drug use may also result in negative effects on brain development.

Direct injury from seizures may also negatively affect the developing brain. Seizures disrupt biochemical/molecular pathways that are normally responsible for activity-dependent development of the maturing nervous system. Seizures can halt or alter cell division, migration, sequential expression of receptor formation, and stabilization of synapses, which contribute to malfunctioning neural networks expressed as neurologic sequelae. Repetitive or prolonged neonatal seizures also increase the susceptibility of the developing brain to suffer subsequent seizure-induced brain injury if epilepsy persists into adolescence and early adulthood. Neonatal animals subjected to status epilepticus have reduced seizure thresholds at later ages as well as impairments of learning, memory, and activity levels when stressed with seizures as adults. Proposed mechanisms of injury also include altered neurogenesis in the hippocampus, possibly due to damage by apoptotic and necrotic pathways. Neonatal seizures may thus initiate a cascade of diverse changes in brain development that become maladaptive at older ages, increasing the risk of damage after subsequent insults.

CLINICAL PEARLS

1. The recognition and classification of neonatal seizures remain problematic. The clinician should optimally rely on continuous synchronized video/EEG/ polygraphic recordings to correlate suspicious behaviors with electrographic seizures.
2. Underlying brain disorders or neonatal diseases encountered after the intrapartum period must be considered as additional or independent causes of neonatal seizures.
3. Treatment choices rely on accurate diagnoses, and either require replacement of glucose, electrolytes, calcium or, alternatively, antiepileptic drugs.
4. Long-term neurologic sequelae following neonatal seizures encompass both epilepsy and comorbid conditions, all of which affect cognition and behavior.
5. More effective treatments are needed that can interrupt the epileptogenic process in the developing brain.

SUGGESTED READING

American College of Obstetricians and Gynecologists' Task Force on Neonatal Encephalopathy and Cerebral Palsy, the American College of Obstetricians and Gynecologists, the American Academy of Pediatrics. *Neonatal Encephalopathy and Cerebral Palsy: Defining the Pathogenesis and Pathophysiology.* Washington, DC: the American College of Obstetricians and Gynecologists, 2003. pp. 1–85.

Holmes GL, Ben-Ari Y. (2001) The neurobiology and consequences of epilepsy in the developing brain. *Pediatr. Res.* 49: 320–325.

Mizrahi EM. (1999) Acute and chronic effects of seizures in the developing brain: lessons from clinical experience. *Epilepsia* 40: S42–S50.

Painter MJ, Scher MS, Alvin J et al.. (1999) A comparison of the efficacy of phenobarbital and phenytoin in the treatment of neonatal seizures. *N. Engl. J. Med.* 341: 485–489.

Scher MS. (2001) Perinatal asphyxia: timing and mechanisms of injury relative to the diagnosis and treatment of neonatal encephalopathy. *Curr. Neurol. Neurosci. Repts.* 1: 175–184.

Scher MS.(2006) Neonatal seizure classification: a fetal perspective concerning childhood epilepsy. *Epilepsy Res.* 70: S41–S57.

Swann JW. (2004) The effects of seizures on the connectivity and circuitry of the developing brain. *MRDD Res. Rev.* 10: 96–100.

Volpe JJ. (2001) *Neurology of the Newborn, 4th ed.*, 178–214. Philadelphia: WB Saunders.

9 Ohtahara Syndrome

W. Donald Shields, M.D.

CONTENTS

CASE PRESENTATION

A 2-month-old girl was transferred from an outside hospital with medically intractable seizures that began on the second day of life. She was the product of an uncomplicated pregnancy, labor, and delivery. However, in retrospect, her mother reports the onset of episodes of occasional, unusually intense fetal movements about two months before delivery. The patient was treated with phenobarbital and did not have seizures for about 1 week. Her seizures then recurred and became increasingly frequent. An EEG was reported to be "abnormal" and an MRI scan was reportedly normal. She had an unremarkable metabolic workup. She received trials of folinic acid and pyridoxine, both of which were unsuccessful in controlling the seizures. The seizures began to increase in frequency and to cluster. She was intubated and treated with IV midazolam for several weeks. The seizures persisted in spite of intense therapy. She had repeated pulmonary infections that became life-threatening, and the midazolam was discontinued. At the time of transfer, she was on phenobarbital and zonisamide.

Examination revealed a well-developed, well-nourished girl who was having frequent, brief, extensor tonic spasms in the awake and sleep states. The general physical exam was unremarkable except for the presence of a gastrostomy tube. On neurologic exam, she appeared to be awake but she did not fix and follow, and had a vacant appearance. She was brachycephalic, diffusely hypotonic, and hyporeflexic, but no other abnormalities were noted. A video/EEG evaluation showed suppression-burst (SB) in the awake and sleep states (Figure 9.1). The extensor spasms were associated

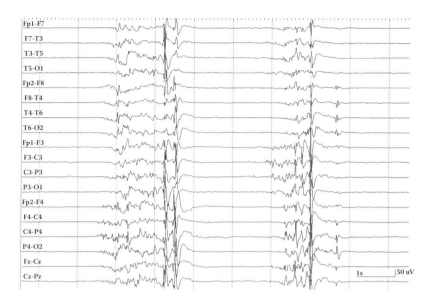

FIGURE 9.1   EEG demonstrating a suppression-burst (SB) pattern.

with attenuation of the SB pattern. A repeat MRI scan was normal. She had a modest improvement with a combination of topiramate and phenobarbital and was able to be discharged home. At five months of age there was a change in the character of her seizures from tonic extensor spasms to flexor spasms. The spasms still occurred in clusters but now were noted to be upon awakening in the morning and after a nap, and no longer in sleep. A repeat EEG showed modified hypsarrhythmia. Adrenocorticotropic hormone (ACTH) was unsuccessful in controlling the infantile spasms. In spite of multiple medication changes, she continued to have occasional infantile spasms until 14 months of age when, again, the character of the seizures gradually changed. The flexor spasms diminished and finally stopped, but were replaced by multiple-seizure types including tonic seizures, atypical absence spells, and generalized tonic–clonic seizures. A repeat EEG showed 2 Hz spike and wave discharges on a slow background. She continued to have occasional seizures on topiramate monotherapy. The patient is now 6-years of age but is developmentally at less than one year.

DIFFERENTIAL DIAGNOSIS

Epileptic encephalopathy that presents in the first few days of life is uncommon. Neonatal seizures are more likely to be due to acute central nervous system (CNS) disturbances such as hypoxic-ischemic encephalopathy, CNS infections, or common metabolic disorders such as hypoglycemia or hypocalcemia. Such disorders are readily diagnosed and managed. Once it is clear that none of the common problems apply, the focus of the differential diagnosis shifts to the less common seizure syndromes. The most important point in the differential diagnosis at this point is to distinguish encephalopathic disorders from benign syndromes such as benign infantile myoclonus or benign neonatal familial convulsions. They are readily differentiated by the clinical course and by an EEG demonstrating a normal background. In our case, it was clear that the patient had a severe form of epilepsy because she had been in prolonged status epilepticus and had failed all attempts at medical therapy, including induction of coma with midazolam. When encephalopathic seizures begin in the first days or weeks of life, the first question to be answered is, what is the epilepsy syndrome? The second question is, what studies are required to determine the underlying cause? Three epileptic encephalopathy syndromes can occur this early in life: early myoclonic encephalopathy (EME); early infantile epileptic encephalopathy (Ohtahara syndrome or EIEE); and infantile spasms (IS or West syndrome). Although there is overlap in the clinical presentation, each of these seizure disorders has a relatively distinct seizure semiology. EME, as the name might indicate, is characterized by myoclonus, which tends to be fragmentary and is typically not associated with an EEG correlate. Patients with Ohtahara syndrome usually have very frequent extensor tonic spasms that may occur dozens or even hundreds of times per day. Tonic spasm may occur concomitantly with the burst, or the SB pattern may be attenuated at the time of the spasm. The more common epileptic encephalopathy syndrome, IS or West syndrome, typically begins later in life, although very early onset has been reported. The spasms of IS are most commonly flexor. They occur in the awake state and tend to cluster. However, some IS patients may have extensor spasms; thus, the presence of extensor spasms is not necessarily an indication of a diagnosis of Ohtahara syndrome. Partial seizures may occur with any of the three syndromes and cannot be used to differentiate between them.

DIAGNOSTIC APPROACH

A careful history and physical/neurologic examination is essential to identify common causes of early onset seizures, as previously noted. Once it is clear that there is no ready explanation for seizures, electroencephalography is the next step in the evaluation, and is very helpful in distinguishing between the many seizures disorders that can present at an early age. Infants who have one of the benign epilepsy syndromes will generally have a normal EEG background rhythm, and may or may not have epileptiform discharges. This is in marked contradistinction to the very abnormal EEG that will always be present in the epileptic encephalopathies. IS is relatively easily distinguished electrographically from the other two encephalopathic syndromes by the presence of hypsarrhythmia, modified hypsarrhythmia, or multifocal independent spike wave discharges. Ohtahara syndrome (EIEE) and EME are both associated with an SB pattern. Thus, in the right clinical setting, an SB EEG would be indicative of either EIEE or EME. It should be noted, however, that SB is not specific to *epileptic* encephalopathy. Neonatal hypoxic ischemic encephalopathy (HIE) is a common cause of SB, and SB can occur in many other disorders. Although the character of the SB is subtly different in EIEE compared with EME, it is very difficult

Ohtahara S, Ishida T, Oka E. et al. (1976) On the specific age-dependent epileptic syndrome: the early-infantile epileptic encephalopathy with suppression burst. *No To Hattatsu.* 8:270–280.

Ohtahara S, Yamatogi Y. (2006) Ohtahara syndrome: With special reference to its developmental aspects for differentiating from early myoclonic encephalopathy. *Epilepsy Res.* 70S: S58–67.

Ohtahara S, Yamatogi Y, Ohtsuka Y. (2008) Ohtahara syndrome. In *Epilepsy: A Comprehensive Textbook 2nd Edition*, Engel J, Pedley T, Ed., 2303–2307. Philadelphia: Lippincott Williams & Wilkins.

Section 3

The Infant

10 Febrile Seizures

Jeffrey R. Buchhalter, M.D., Ph.D.

CONTENTS

CASE PRESENTATION

Scenario 1: A 12-month-old child has been irritable and lethargic for the last 12 hours. He feels warm to the touch and has been placed in lukewarm water baths. Suddenly, he stiffens and then becomes limp with fine shaking of all extremities for less than 1 minute. Within approximately 15 minutes, he returns to his pre-seizure status.

Scenario 2: Same scenario as above except that the shaking involves the left arm and leg, and persists for 20 minutes.

DIFFERENTIAL DIAGNOSIS

A febrile seizure (FS) is usually defined as a seizure related to fever (often defined as a temperature >38.4°C) in a child between 1 to 6 months and up to 5 years of age, and in the absence of an intracranial infection. There is some variation in the literature with regard to the minimum and maximum ages, degree of temperature elevation, and requirement for neurological normalcy. A *complex* febrile seizure (CFS) is distinguished from a *simple* FS (SFS) by a seizure duration of greater than 10–15 minutes, symptoms or signs of focality, and two or more events within 24 hours. Febrile status epilepticus may be seen in about 5% of patients with febrile seizures.

The differential diagnosis of any type of potential seizure should include consideration of relevant paroxysmal, nonepileptic events. In the age range considered for FSs, possibilities include rigors associated with illness, gastroesophageal reflux, and breathholding spells. The latter entities usually do not occur in the context of fever but certainly could do so either by coincidence or with fever as a provoking factor. Recently, the entity of afebrile seizures occurring in Asian and Caucasian children in the context of gastroenteritis has been reported with a prognosis similar to that of FSs. However, the critical differential diagnostic entities are meningitis, encephalitis, and cerebral abscess, as these are potentially life-threatening disorders.

DIAGNOSTIC APPROACH

A detailed history and physical examination will usually serve to rule out the paroxysmal nonepileptic events described previously that can occur during a febrile illness. Diagnostic confusion exists when the individual who brings the child to medical attention was not the person who actually witnessed the event. There is no evidence to support the use of obtaining hematological or other biochemical testing as part of the routine evaluation of an uncomplicated FS.

The history of a child with a prolonged illness, progressive encephalopathy, and physical findings—including a bulging fontanelle, meningismus, and tonic posturing—raises the specter of seizures secondary to an intracranial infection. Due to the lack of specificity of symptoms and signs, the child younger than 18 months deserves special consideration. In this circumstance, a lumbar puncture (LP) should be considered, especially if no contraindications such as a bleeding diathesis and/or herniation exist. Potential herniation is suggested by fundoscopic abnormalities, asymmetric pupils, and posturing.

Brain computed axial tomography (CAT) will not necessarily rule out increased intracranial pressure but will serve to evaluate for the presence of space-occupying mass lesions (e.g., tumor, abscess). It should be emphasized that the acutely ill child with an SFS who is interactive and has a normal neurological examination is most unlikely to have bacterial meningitis or warrant CT imaging. Magnetic resonance imaging (MRI) can be performed on an elective basis if an underlying structural abnormality is suggested by focal seizures, focal signs, or static or progressive encephalopathy.

Surprisingly little data exist regarding the diagnostic yield of an electroencephalogram (EEG) following an FS. In one retrospective, uncontrolled study of 33 patients who experienced an FS, no significant abnormalities were found in EEGs performed within one week of the seizure. Of note, 24 of the children had complex FSs, which perhaps influenced the decision to obtain the study. Also, EEG does not appear to be predictive for the risk of recurrence of febrile seizures or later epilepsy. Therefore, an EEG is not routinely recommended for patients with SFSs, and may be of limited utility in patients with complex febrile seizures.

TREATMENT STRATEGY

The child who has experienced a FS is usually seen in an acute care setting such as an emergency department or in the primary care clinic, having returned to his or her baseline level of functioning. The issues at that time include ascertainment of the source of fever/infection and control of fever with antipyretics. Thereafter, one can consider available treatment modalities to prevent recurrence in the future, which include symptomatic treatment of febrile episodes with an antipyretic agent and/or a rapidly acting benzodiazepine and chronic, prophylactic antiepileptic drugs.

Several randomized, double-blind studies have evaluated antipyretics with or without an AED to determine if recurrence rates could be decreased. No impact of antipyretic treatment could be demonstrated, acknowledging that the studies had significantly different study populations and designs. The lack of effect was present in a study that selected children on the basis of having at least one risk factor for recurrence, thereby enhancing the likelihood of recurrence in this group. In the child with one or few SFSs, an antipyretic can be recommended only to make the child more comfortable during the febrile illness. The family should be informed that there is no evidence to suggest that this measure will decrease the likelihood of another event.

The use of benzodiazepines has been studied for the treatment and prevention of febrile seizures. Oral or rectal diazepam at the onset of a febrile illness has been shown to decrease the recurrence of febrile seizures compared to placebo. The sedative side effects of the medication limit the utility of this option in most instances. Rectal diazepam or diazepam gel can be used as abortive therapy for the child who is subject to recurrent or prolonged (complex) FSs to potentially decrease the duration of the ongoing seizures. A specific—albeit arbitrary—duration (such as 5 minutes) should be decided upon with the family, at which time a benzodiazepine is administered. The willingness and ability of the family to deal with this more extreme form of FS should be taken into account.

Finally, the use of a chronic, prophylactic AED should be used only if the perceived benefit outweighs the known dose-related and idiosyncratic adverse effects. Although there is some evidence that prophylaxis with phenobarbital and valproic acid reduces recurrence, there are no data that indicate a protective effect against the later development of epilepsy. There has been little to no research into the chronic daily use of the newer AEDs in the prevention of febrile seizure recurrence. Currently, there are no pediatric professional organizations that include AED prophylaxis as part of a practice guideline for FSs.

LONG-TERM OUTCOME

There are no data to convincingly suggest that single or multiple, simple or complex FSs are associated with *cognitive, behavioral,* or *motor morbidities.* The exception is found in those relatively few children who go on to develop a medically refractory seizure disorder. Similarly, there does not appear to be any increased risk of premature *mortality* in children with FSs, a fact that should be shared with concerned families.

A large body of literature relates to the risk factors that predict *recurrence* of FSs and developing *afebrile seizures* (i.e., epilepsy). In summary, approximately one-third of children will have a second FS following the initial event. An increased risk of recurrence is found in those individuals who have the first FS at 18 months of age or less, have a first-degree family member with FS, have multiple FS within 24 hours, and an occurrence of the FS at a relatively low temperature and within 1 hour of illness onset.

Although the overall risk of having an afebrile seizure following a FS is low (2–4% in Western Europe and North America), families should not be told that FSs always have a benign long-term outcome despite the prognosis being excellent for the vast majority of individuals. Risk factors associated with an increased risk of subsequent afebrile seizures includes a family history of epilepsy, complex features with the FS, and any indication of nervous system abnormalities (e.g., structural, motor, cognitive). One population-based study reported that almost 50% of children will develop epilepsy if followed for 25 years and if three complex features are present.

The relationship between FS and the later development of mesial temporal lobe epilepsy remains controversial. The reported association is strongest in retrospective case series from epilepsy surgery centers, whereas population-based epidemiologic studies do not support such a relationship. However, several reports have documented hippocampal sclerosis following prolonged FSs. This finding has not been replicated in prospective neuroimaging studies of children with FS.

NEUROBIOLOGY/PATHOPHYSIOLOGY OF DISEASE

The cause or causes of simple and complex febrile seizures are unknown, but are likely to be both multifactorial and developmental in nature. The precise role of fever is uncertain, and there is no evidence to support the impression that the seizure is related to the rate of temperature rise. The role of cytokines such as *Il-1* has been hypothesized, but it is unclear whether alterations in this biochemical, inflammatory response system is causal or secondary. The importance of genetic factors is suggested by epidemiological data indicating increased risk of occurrence in families with a known history of FSs. Furthermore, a number of genes, predominantly involving voltage-dependent sodium channels, have been associated in family clusters with FSs and the syndrome of generalized epilepsy with febrile seizures plus (GEFS+), as described elsewhere in this volume.

CLINICAL PEARLS

1. The major differential diagnosis regarding febrile seizures is to rule out an intracranial infection.
2. Approximately one-third of children with an initial FS will have a second, and the likelihood is determined by a set of well-defined risk factors.
3. Epilepsy develops in a small proportion of children following the initial FS, with the likelihood related to the number of complex features present and the family history of epilepsy.
4. There are no data to support use of routine blood work, neuroimaging, lumbar puncture, or EEG following a SFS.

SUGGESTED READING

Annegers JF, Hauser WA, Shirts SB, Kurland LT. (1987). Factors prognostic of unprovoked seizures after febrile convulsions. *New Eng. J. Med.* 316: 493–498.

Carroll W, Brookfield D. (2002). Lumbar puncture following febrile convulsion [see comment]. *Arch. Dis. in Childhood* 87: 238–240.

Maytal J, Steele R, Eviatar L, Novak G. (2000). The value of early postictal EEG in children with complex febrile seizures. *Epilepsia* 41: 219–221.

Offringa M, Moyer VA. (2001). An evidence-based approach to managing seizures associated with fever in children. *West. J. Med.* 175: 254–259.

Purssell E. (2000). The use of antipyretic medications in the prevention of febrile convulsions in children. *J. Clinical Nursing* 9: 473–480.

Rantala H, Tarkka R, Uhari M. (1997). A meta-analytic review of the preventive treatment of recurrences of febrile seizures [see comment]. *J. Pediatrics* 131: 922–925.

Waruiru C, Appleton R. (2004). Febrile seizures: an update. *Arch. Dis. Childhood* 89: 751–756.

11 Generalized Epilepsy with Febrile Seizures Plus (GEFS+)

Noel Baker, M.D.

CONTENTS

CASE PRESENTATION

The index patient is a 9-year-old girl with a history of approximately 30 generalized tonic–clonic seizures beginning at 5 months of age. The seizures initially began with fever, but later occurred spontaneously in the absence of fever or intercurrent illness. From 3 to 4 years of age, she experienced absence seizures. An initial electroencephalogram (EEG) demonstrated generalized 3 Hz spike/wave discharges. She was given the diagnosis of childhood absence seizures, and was treated with lamotrigine. Despite adequate doses, she continued to have occasional seizures. A repeat EEG was interpreted as normal. Lamotrigine was discontinued in lieu of valproic acid, which made her seizure free. All along, her neurological examination was normal. Interestingly, her family history was positive for seizures in her older brother, who experienced a total of 50 generalized tonic–clonic seizures with and without fever beginning at age 7 months. He also had generalized myoclonic seizures from age 7 months to 3 years. An EEG at age 4 years was notable for mild background slowing, but similar to his younger sister, a repeat study was normal (see Figure 11.1), as was his neurological exam. Notably, other family members experienced seizures with and without fever, including another brother (from 1 to 25 years of age), a first cousin, and an aunt. One of the aunt's sons was

localization-related, epilepsy syndromes. Vagus nerve stimulation or the ketogenic diet may be helpful in selected medically refractory cases. Resective epilepsy surgery is generally not indicated in patients, but rare individuals with hippocampal sclerosis have been reported in GEFS+ families.

LONG-TERM OUTCOME

Long-term outcome depends upon seizure type or types, initial presentation, and whether the neurologic examination is normal. Importantly, the specific genetic abnormality may be a critical determinant of the clinical outcome, but no clear genotype/phenotype correlations have been established to date. Prognosis appears to be related to the actual epilepsy syndrome that the patient exhibits. In the original 1997 report describing a large extended GEFS+ family, most members had benign and self-limited forms of the syndrome, such as febrile seizures persisting beyond the usual age, or febrile seizures with absences. However, one individual had generalized seizures that persisted into middle age, and another had severe refractory MAE. As noted previously, SMEI is not a benign disorder, and the majority of such patients have significant cognitive impairment and persistent epilepsy.

PATHOPHYSIOLOGY/NEUROBIOLOGY

Genetic mutations causing a loss or gain of function are presumed in all cases of GEFS+. Some studies suggest that a mutation in the gamma-2 subunit of the $GABA_A$ receptor leads to a decrease in the inhibitory GABA-induced currents, thus leading to neuronal hyperexcitability. Research involving mutations in genes encoding various subunits of the voltage-gated sodium channel indicates that changes in sodium channel function can range from net increases in inward sodium current (via slowing of inactivation or reduction in the sodium current rundown, seen with *SCN1A* mutations) to a shortening of the refractory period following an action potential (with *SCN1B* mutations). Although there remains an absence of clear genotype/phenotype correlations, many of the more benign GEFS+ patients were found to have missense mutations, whereas the SMEI phenotype was more often associated with truncation defects. Unidentified susceptibilty genes probably account for some of the variability and incomplete penetrance seen in this disorder. This is an evolving area of research, and at present, genotype is not predictive enough of phenotype to be clinically useful. Nevertheless, genetic counseling should be performed in all patients and their families with proven mutations.

CLINICAL PEARLS

1. GEFS+ is a heterogeneous genetic syndrome most often associated with febrile seizures.
2. The diagnosis is suggested on the basis of a strong family history of febrile and afebrile seizures and/or generalized epilepsy.

3. Given the broad spectrum of disease, treatment for GEFS+ should be based on features of the specific epilepsy syndrome affecting the individual.

4. Genetic mutations in voltage-gated sodium channels and subunits of $GABA_A$ receptors have been associated with GEFS+, but the specific type of mutation cannot accurately predict prognosis or response to treatment.

SUGGESTED READING

Audenaert D, Van Broeckhoven C, De Jonghe P. (2006). Genes and loci involved in febrile seizures and related epilepsy syndromes. *Hum. Mutation* 27(5): 391–401.

Baulac M, Gourfinkel-An I, Baulac S, Leguern E. (2005). Myoclonic seizures in the context of generalized epilepsy with febrile seizures plus (GEFS+). *Advs. Neurology* 95: 103–125.

Nakayama J, and Arinami T. (2006). Molecular genetics of febrile seizures. *Epilepsy Res.* 70 Supplement 1: 190–198.

Scheffer ID, Berkovic SF. (1997). Generalized epilepsy with febrile seizures plus: a genetic disorder with heterogeneous clinical phenotypes. *Brain* 120: 479–490.

12 Benign Myoclonic Epilepsy of Infancy

Kristen L. Park, M.D. and
Douglas R. Nordli, Jr., M.D.

CONTENTS

CASE PRESENTATION

A 10-month-old boy presented with intermittent body jerks. These had been noted by his parents for the preceding three weeks, but have become more intense. The jerks occur when the infant is awake and are characterized by sudden flexion of the neck, shrugging of the shoulders, and elevation of the arms. They may occur singly or in brief clusters of two to three repetitive jerks with a repetition rate faster than 3 Hz. The jerks have not resulted in any drops, and the child quickly resumes activities without any apparent alteration. There have been no other seizures. This occurs against the backdrop of a normal child. There are no risk factors for epilepsy and no family members with seizure disorders or other neurological issues. General physical examination and detailed neurological examination were both normal. Because the jerks were occurring with increasing frequency, 2–3 hours of video-EEG was performed to capture and better characterize the events. Several typical jerks were recorded. The clinical events consisted of a brief head nod with elevation of the shoulders and arms. The EEG accompaniment was a brief burst of diffuse spike–wave discharges (Figure 12.1). After extensive discussion of treatment options, the family declined treatment with valproic acid and elected to begin levetiracetam. Within two weeks, the boy developed a diffuse erythematous rash involving the palms and soles.

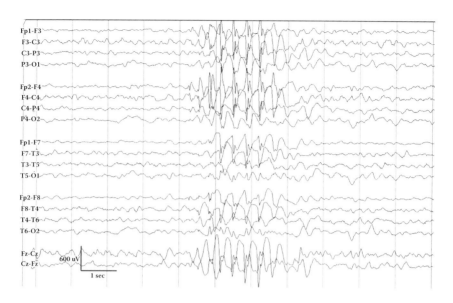

FIGURE 12.1   EEG tracing from the patient during drowsiness, sensitivity 30 uV/mm. The generalized spike–wave discharges correlated with several repetitive jerks of both arms.

The mucous membranes were spared. An environmental allergen was suspected as the family was doing construction in the home; however, a hypersensitivity reaction to the medication could not be excluded. The medication was discontinued and the child received corticosteroids for 5 days. His parents noted a complete cessation of the jerks. The child was seen for regular follow-up 1 year later without return of seizures. His development has remained normal during this time.

DIFFERENTIAL DIAGNOSIS

The first consideration when approaching this case should be determining whether the jerks are epileptic in nature. Some nonepileptic conditions that mimic myoclonic attacks can often be distinguished from BMEI by a careful history that focuses on time course and a careful description of the events. Hypnic myoclonus can be excluded if attacks are witnessed during wakefulness. Shuddering attacks may be seen in children as young as 4–6 months of age, but are usually characterized by stiffening of the body, adduction of the knees, flexion or extension of the neck, and high-frequency trembling lasting several seconds. In these children there may also be a family history of essential tremor. Nonepileptic myoclonus can also be seen in association with opsoclonus; however, in this case, its onset is usually progressive and follows that of the abnormal eye movements. Other types of nonepileptic myoclonus, such as those seen in a variety of metabolic and genetic diseases, can be excluded by a normal neurologic examination and developmental assessment.

Once these conditions have been excluded, an EEG should be obtained to examine the background activity and capture an event so that it can be correlated with any ictal discharge. Capturing an event in this instance is critical as infants with BMEI, and other entities within the differential diagnosis, may have normal interictal EEGs, at least initially. If there is no EEG correlate, benign infantile myoclonus is the likely diagnosis as it is nonepileptic in origin. Infantile spasms may be confused with myoclonic events but can usually be distinguished by several important clinical and electrographic features. Spasms usually consist of a more prolonged tonic contracture rather than the lightening jerk of myoclonus, are more often seen upon awakening rather than drowsiness, and are usually accompanied by hypsarrhythmia and elecrodecrements on EEG examination. Severe myoclonic epilepsy of infancy (Dravet syndrome) should also be considered, but often presents with prolonged febrile convulsions followed by afebrile focal seizures. Although the interictal EEG may be initially normal, it usually develops epileptiform discharges and slowing of the background rhythms. Symptomatic generalized epilepsy (Lennox–Gastaut syndrome, for example) may rarely present in this age group with prominent myoclonia but was considered unlikely in light of the child's normal development, EEG background, and neurological examination.

DIAGNOSTIC APPROACH

As discussed in the previous section, a careful history and neurologic examination are essential for the diagnosis of this syndrome. The most useful supplement to these clinical tools is the video-EEG, which will help classify the myoclonus as epileptic and rule out other epilepsy syndromes. The background of the EEG is often normal, and interictal abnormalities are uncommon. The myoclonus is associated with generalized spike-wave or polyspike-wave discharges usually lasting 1–3 seconds. In some patients, the myoclonus is stimulated by intermittent photic stimulation, whereas reflex myoclonus may be triggered by a loud sound or other stimulus. Neuroimaging is not indicated in typical cases as this is an idiopathic syndrome not associated with structural malformations.

TREATMENT STRATEGY

The first question to consider once the correct diagnosis has been made is whether or not treatment is indicated. The clinician and family must balance the side effects of medication against the impact of ongoing seizures on the developing brain. Unfortunately, the first of these two is more apparent and quantifiable than the latter. Given the available evidence, treatment should be offered to families with the knowledge that its ultimate effect on outcome is uncertain. Based on limited case series and review of the literature, some authors suggest that an earlier onset of seizures and delay of treatment may be associated with poorer outcome. However, this association has not been borne out in other studies, and there is insufficient evidence to conclude that latency to treatment and age of onset are independent predictors of outcome, especially given the broad underlying spectrum of the condition itself.

Although there is no evidence from randomized controlled trials, the first line agent for treatment is considered to be valproic acid (VPA). Seizure freedom rates using VPA range from 77–95% of patients. Other treatments have included benzodiazepines, phenobarbital, lamotrigine, ethosuximide and, more recently, levetiracetam. A refractory case is cited in which combination therapy and the ketogenic diet failed to result in remission of seizures. Given this natural history, adequate monotherapy trials with monitoring of blood levels should be documented before additional agents are added. Routine follow-up should be scheduled to monitor the child's development and to screen for other seizure types, regardless of whether treatment is pursued.

Although isolated reflex myoclonic seizures do not represent a distinct syndrome, the cognitive outcome may be more benign. Therefore, these seizures are often not treated with antiepileptic medication.

There is no current consensus regarding the length of treatment for BMEI. Most published literature recommends a treatment course of several years, after which medication can be tapered if the patient has remained seizure free. Patients with isolated reflex myoclonic seizures may avoid treatment altogether or be tapered after a shorter seizure-free interval of 1 year.

LONG-TERM OUTCOME

Recent reviews have suggested that the cognitive outcome for these children is not as benign as originally reported. The incidence of neuropsychiatric disturbance in these children ranges from 21–58%. Cognitive problems ranging from learning disability to mental retardation were seen. In addition to cognitive disturbances, other neurologic abnormalities have been reported, including fine motor deficits, hyperkinesias, attention deficits, and behavioral problems. Ten reported cases of purely reflex myoclonic epilepsy in one review had normal outcomes.

The prognosis for long-term seizure control is more encouraging with remission seen in the vast majority of cases. Recurrence of seizures has been reported with a frequency of approximately 18% of published cases, and most often consists of generalized tonic–clonic seizures. In patients with marked photosensitivity, the recurrence risk seems to be higher, and more prolonged treatment may therefore be advisable.

PATHOPHYSIOLOGY/NEUROBIOLOGY OF DISEASE

The underlying basis of this disorder is not known, and it is currently classified among the idiopathic generalized epilepsies. There is most likely some genetic component, as approximately 50% of cases report a family history of epilepsy or febrile seizures. Familial cases of infantile myoclonic epilepsy with autosomal recessive inheritance have been identified and linked to chromosome 16p13. Candidate genes in this region include voltage-dependent chloride channel 7, synaptogyrin III, and the solute carrier family 9 isoform 3 regulatory factor 2 gene.

Although these cases differ in several regards from the syndrome defined by Dravet—most notably the presence of other seizure types and a more severe phenotype—it has been suggested that BMEI and other myoclonic epilepsies of infancy may represent a spectrum with variable expression and severity. In addition, genetic linkage studies of large affected families have begun to identify candidate genes for other epilepsies in this category, including childhood absence epilepsy, severe myoclonic epilepsy of infancy (Dravet syndrome), and generalized epilepsy with febrile seizures plus. In addition to genetic factors, brain maturation also plays a role in the timing of this disorder, which peaks between 6 months and 3 years of age. As such, it has some superficial resemblance to other age-related epileptic phenomenon (benign neonatal convulsions, febrile seizures, infantile spasms, etc.) that are unique to the developing brain.

CLINICAL PEARLS

1. Benign myoclonic epilepsy of infancy should be considered when a normally developing child between the age of 6 months and 3 years presents with myoclonus during wakefulness.
2. A family history of seizures is seen in an estimated 30–50% of patients, and children may often have a preceding history of febrile seizures.
3. The EEG demonstrates generalized spike–wave discharges associated with the myoclonias, but the background is usually normal.
4. Treatment is usually, but not invariably, recommended, and the first-line agent is valproic acid.
5. Despite its name, this disorder can be associated with cognitive impairment, neuropsychiatric disturbances, and seizures that require ongoing treatment.

SUGGESTED READING

Auvin, S., Lamblin, M.D., Pandit, F., Bastos, M., Derambure, P., Vallée, L. (2006) Benign myoclonic epilepsy in infants: electroclinical features and long-term follow-up of 34 patients. *Epilepsia.* 47, 387–93.

Dravet, C., Bureau, M. (2005) Benign myoclonic epilepsy in infancy. *Adv. Neurol.* 95, 127–37.

Lombroso, C. T., Fejerman, N. (1977) Benign myoclonus of early infancy. *Ann. Neurol.* 1, 138–43.

Mangano, S., Fontana, A., Cusumano, L. (2005) Benign myoclonic epilepsy in infancy: neuropsychological and behavioural outcome. *Brain Dev.* 27, 218–23.

Ricci, S., Cusmai, R., Fusco, L., Vigevano, F. (1995) Reflex myoclonic epilepsy in infancy: a new age-dependent idiopathic epileptic syndrome related to startle reaction. *Epilepsia.* 36, 342–8.

Zara, F., Gennaro, E., Stabile, M., Carbone, I., Malacarne, M., Majello, L., Santangelo, R., de Falco, F.A., Bricarelli, F.D. (2000) Mapping of a locus for a familial autosomal recessive idiopathic myoclonic epilepsy of infancy to chromosome 16p13. *Am. J. Hum. Genet.* 66, 1552–7.

Zuberi, S. M., O'Regan, M. E. (2006) Developmental outcome in benign myoclonic epilepsy in infancy and reflex myoclonic epilepsy in infancy: a literature review and six new cases. *Epilepsy Res.* 70 Suppl. 1, S110–5.

13 Severe Myoclonic Epilepsy in Infancy

Matthew M. Troester, D.O.

CONTENTS

CASE PRESENTATION

The patient was born at term, without complications, to a 28-year-old female after an uneventful pregnancy. Her birthweight was 3.1 kg, and she was discharged at the same time as her mother. She met all of her early developmental milestones and was beginning to crawl, when at six months of age she became lethargic and was found to have a temperature of 38.5°C. She was noted to have jerking of one side of her body, which then spread to involve her entire body lasting less than 5 minutes. She was taken to the local children's hospital where an electroencephalogram (EEG) and magnetic resonance imaging (MRI) scan of her brain were performed. Her mother was told these were normal and that her daughter had a complex febrile seizure (FS). There was no family history of epilepsy, with or without fever. The patient did crawl, but her development plateaued at cruising around furniture. Her speech never developed beyond two to three words. She continued to have seizures associated with fever. She followed up with the neurologist who repeated the previous studies, finding a similar MRI result but an EEG with very frequent, multifocal, independent spike and wave discharges, which at times appeared irregularly generalized. She underwent an extensive evaluation amino acids, including chromosomal microarray analysis, lysosomal enzymes, serum amino acids, urine (copper, amino acids, organic acids), spinal fluid (lactate, biogenic amines), nerve (nerve conduction velocity), and muscle (biopsy for histology, electron microscopy, and mitochondrial enzymes).

All of these tests were unrevealing, as was a dilated funduscopic examination by a pediatric ophthalmologist. The patient started to have more frequent seizures without fever and developed multiple seizure types including absence and myoclonic seizures. A third EEG revealed abundant multifocal epileptiform discharges. She was referred for DNA sequencing of the neuronal voltage-gated sodium channel alpha 1 subunit (*SCN1A*) gene, and a two base-pair deletion of uncertain significance was found. Further testing of her asymptomatic parents failed to reveal evidence of a similar mutation, and a diagnosis of severe myoclonic epilepsy of infancy (SMEI) was made.

DIFFERENTIAL DIAGNOSIS

Alternative diagnoses to SMEI include: complex febrile seizures (CFS), myoclonic astatic epilepsy (Doose syndrome), benign myoclonic epilepsy, progressive myoclonic epilepsy (PME), Lennox–Gastaut syndrome, and cryptogenic localization-related epilepsy.

Early confusion between Dravet syndrome and CFSs is common. FSs more commonly have their onset after the first year of life, whereas Dravet syndrome starts earlier (between 2 and 12 months of age). Dravet syndrome is characterized by unilateral seizure activity in association with modest temperature elevation (at or less than 38°C), and prolonged seizures in spite of proper treatment. Children with FSs (even complex events) have consistently normal development and unremarkable EEGs. Although the early EEG in Dravet syndrome may be normal, later tracings reveal generalized, focal, or multifocal epileptiform discharges. No single pattern is specific for Dravet syndrome, and photosensitivity is a variable finding. Other epilepsy syndromes can present with early FSs, including myoclonic astatic epilepsy (Doose syndrome), and benign myoclonic epilepsy (BME), the latter rarely. Doose syndrome patients have recurrent drop attacks as the predominant seizure type and more generalized EEG findings. BME patients have only myoclonic seizures and normal development. The progressive myoclonic epilepsies can be confused with Dravet syndrome but typically have a more degenerative course and can be associated with ophthalmologic disturbances. Abnormalities are often found upon examination of spinal fluid and muscle in PME, depending on the etiology. Some patients with cryptogenic localization-related epilepsy will present with seizure and fever in infancy but do not go on to develop various predictable seizure types or consistent developmental impairments. Lennox–Gastaut syndrome (LGS) shares some seizure types with SMEI: However, LGS features drop attacks as a major seizure type, appears later in childhood, and has a distinct interictal slow spike-and-wave EEG discharge pattern.

DIAGNOSTIC APPROACH

The name "severe myoclonic epilepsy of infancy" can be misleading, as myoclonic seizures are not often the presenting or most notable seizure type. The International League against Epilepsy denotes Dravet syndrome as the infant who usually presents with prolonged, unilateral, and/or generalized seizures in the setting of mildly elevated temperature, and then progresses to other seizure types and developmental impairment.

Other seizure types seen in Dravet syndrome include myoclonias, partial seizures, and atypical absences. These occur starting in the second or third year of life and can accompany the early seizures associated with fever. Seizure flurries are not uncommon, and episodes of convulsive and nonconvulsive status epilepticus occur. Myoclonic seizures variably affect axial musculature and cause violent drops or barely perceptible head nods. Subtle myoclonias are also noted as brisk jerks of the extremities and tend to abate with time. Partial seizures, both simple and complex, also occur and can generalize. Atypical absences may occur and may be associated with a myoclonic jerk or nod.

Intellectual stagnation coincides with the onset of these other seizure types. Most authors describe a static encephalopathy, though regression does not exclude the diagnosis of Dravet syndrome. Language and walking typically occur on time but do not develop normally. More than half the patients with Dravet syndrome will develop variable degrees of ataxia, and about 20% develop subtle pyramidial signs.

The EEG lacks a specific signature or feature, and the MRI findings in Dravet are equally noncharacteristic. Nevertheless, a Hungarian series noted the occurrence of hippocampal sclerosis with serial studies through the course of the disease in a majority of patients who had normal MRIs at disease onset.

Fortunately, Dravet syndrome is uncommon, with an estimated incidence of less than 1 in 20,000 to as few as 1 in 40,000, depending on the author. Many patients will have a family history of FSs or epilepsy. Nearly 75% of patients with a Dravet or Dravet-like phenotype will exhibit *SCN1A* gene mutations. These Dravet-like phenotypes have also been referred to as borderland SMEI syndromes (SMEIB) because, whereas these patients have seizures and encephalopathy, they lack one or two typical clinical or EEG findings such as generalized spike wave activity. Dravet syndrome exists on a continuum with other severe epilepsies of infancy and is on the opposite side of this continuum from generalized epilepsy with febrile seizures plus (GEFS+), a milder phenotype of *SCN1A* gene mutations.

TREATMENT STRATEGY

Any attempt to limit exposure to febrile illness is helpful (i.e., avoid daycare-type settings). Vaccinations are not contraindicated. Scheduled antipyretics and extra doses of antiepileptic medications around the time of vaccination may be useful. A properly fitted helmet is advised for those patients who fall from their myoclonic seizures. Providing families with abortive therapy, such as rectal diazepam, may allow for earlier treatment for episodes of status epilepticus.

Opinions regarding efficacious medications vary, and no single agent stands out. Treatment with valproic acid along with a benzodiazepine is commonly initially advocated. Stiripentol, vigabatrin, and topiramate have demonstrated some efficacy in limited trials or reports. Some authors prefer clonazepam to clobazam. Newer trials with much older agents such as potassium bromide have shown promise against convulsive episodes. Other drugs such as carbamazepine and lamotrigine can aggravate seizures. The ketogenic diet may help some patients. Immunomodulation with corticosteroids and immunoglobulins have limited efficacy and create further concerns about susceptibility to infection.

LONG-TERM OUTCOME

Although a majority of patients survive childhood, patients with Dravet syndrome are significantly intellectually impaired and continue to have treatment-resistant seizures, some still associated with fever. Myoclonic and atypical absence seizures may abate after childhood, and partial seizures are less common in older patients.

PATHOPHYSIOLOGY/NEUROBIOLOGY OF DISEASE

Voltage-gated sodium channels facilitate initiation and propagation of action potentials. Their activation causes the initial upstroke of the action potential by allowing a few positive sodium ions into the cell to reverse the normally negative potential inside the cell. These sodium channels then rapidly close, and potassium channels open to initiate repolarization. This sequence triggers events such as neuronal firing or muscle contraction. Mutations in the genes encoding human sodium channel alpha subunits result in structural alterations that, through gain or loss of function, lead to seizure propensity. Over 100 mutations have been described and most arise de novo.

CLINICAL PEARLS

1. Dravet syndrome usually presents with repeated and prolonged FSs in the first year of life.
2. Multiple seizure types may be seen, including generalized tonic–clonic, myoclonic, complex partial, and atypical absence seizures.
3. Seizures typically remain resistant to medical treatment, but may decrease with age.
4. A family history of epilepsy is often seen, and mutations in the *SCN1A* gene may be found in up to 75% of patients.

SUGGESTED READING

Arzimanoglou A, Guerrini R, Aicardi J. (2004). Dravet Syndrome: Severe myoclonic epilepsy or severe polymorphic epilepsy of infants. In *Aicardi's Epilepsy in Children, 3rd Edition*, 51–57. Philadelphia: Lippincott Williams & Wilkins.

Dravet C, Bureau M, Oguni H, Fukuyama Y, Cokar, O. (2005). Severe myoclonic epilepsy in infancy: Dravet Syndrome. *Adv. Neurology* 95, 71–102.

Harkin L, McMahon J, Iona X. et al., Infantile Epileptic Encephalopathy Referral Consortium, Sutherland, G., Berkovic, S., Mulley, J., Scheffer, I. (2007). The spectrum of SCN1A-related infantile epileptic encephalopathies. *Brain* 130: 843–852.

Siegler Z, Barsi P, Neuwirth M. et al. (2005). Hippocampal sclerosis in severe myoclonic epilepsy in infancy: a retrospective MRI study. *Epilepsia* 46(5): 704–708.

14 Infantile Spasms

Richard A. Hrachovy, M.D. and
James D. Frost, Jr., M.D.

CONTENTS

CASE PRESENTATION

The patient was an 8-month-old white male delivered by Cesarean section at 36 weeks gestational age. His birthweight was 7 pounds and 6 ounces. No complications were noted during pregnancy, and he was discharged home shortly after birth. There was no family history of seizures, and the patient had a 5-year-old sister who was healthy. The patient's parents first noted unusual spells at 6 months of age. The spells consisted of stiffening of the extremities and abduction of the arms, and these events occurred in clusters several times throughout the day. The clusters lasted from 5–10 minutes, and the patient often cried following the stiffening episodes. Occasionally, the patient was noted as looking to the right prior to a cluster. No other seizures were noted. Developmentally, the patient could not support his head and could not sit without support. He did not cruise. He made sounds but spoke no discernable words. On physical examination, his occipital-frontal head circumference was 45 cm (50th percentile), and his length was 62 cm. His general physical examination was unremarkable except for the skin. A single hypopigmented lesion was found on the patient's back, using a Wood's lamp. His neurological examination was notable for decreased truncal and lower extremity tone, and he could not sit without support.

A brain magnetic resonance imaging (MRI) scan revealed multiple cortical tubers and numerous subependymal nodules along the margin of the lateral ventricles, findings consistent with tuberous sclerosis. An electroencephalogram

(EEG) was obtained, which revealed a hypsarrhythmic pattern. In addition, during the EEG, a cluster of epileptic spasms was recorded. Each spasm was associated with a generalized sharp-wave transient followed by a generalized voltage attenuation. A diagnosis of infantile spasms was made based upon the history and EEG findings.

DIFFERENTIAL DIAGNOSIS

The epileptic spasms associated with this disorder typically occur as brief, symmetrical contractions of the musculature of the neck, trunk, and extremities. The intensity and pattern of distribution of the spasms is highly variable, and three main types of motor spasms may occur: flexor, extensor, and mixed flexor–extensor. Less commonly, asymmetrical spasms may be seen, and periods of behavioral arrest may occur following a motor spasm. The ictal events may occur in isolation, but most frequently occur in clusters. The spasms may occur throughout the day and night, but rarely occur during sleep. Instead, they frequently occur immediately upon arousal from sleep.

Infantile spasms may be confused with a variety of normal and abnormal infant behaviors. The brief, transitory nature of epileptic spasms makes it difficult for parents or other observers to provide an accurate description of the episodes, and consequently, the diagnosis of infantile spasms is often delayed for weeks or months because parents and physicians do not recognize the motor phenomena as seizures. Parents may confuse spasms with Moro reflexes, other startle responses, hypnagogic jerks occurring during sleep, head-banging, and transient flexor–extensor posturing of trunk and extremities of nonepileptic origin. They may also be confused with other nonepileptic medical conditions such as spasmus nutans, colic, benign myoclonus of early infancy, and benign neonatal sleep myoclonus.

Several epileptic syndromes of infancy and early childhood may be confused with infantile spasms. These include benign myoclonic epilepsy in infancy, severe myoclonic epilepsy in infancy, epilepsy with myoclonic-astatic seizures, early infantile epileptic encephalopathy (EIEE, or Ohtahara syndrome), early myoclonic encephalopathy (EME), and Lennox–Gastaut syndrome. The latter three syndromes share many characteristics with infantile spasms and are typically separated from each other primarily by age of onset. The fact that these syndromes may transition from one to the other also complicates the issue.

The differentiation of infantile spasms from nonepileptic events, other types of myoclonic activity, and other epileptic syndromes usually requires continuous video-EEG monitoring to provide a definitive diagnosis.

DIAGNOSTIC APPROACH

The diagnosis of infantile spasms is suggested on the basis of the clinical history, especially the description of spasm-like events that occur in clusters on arousal from sleep. Meticulous general and neurological examinations must be performed, including a careful ophthalmologic evaluation and close examination of the skin with a Wood's lamp to search for such conditions as tuberous sclerosis (i.e., hypopigmented macules). A routine EEG, recorded with the patient awake and asleep, should then be obtained to help establish the diagnosis. If the routine EEG does not show hypsarrhythmia, and if the events in question are not recorded, a prolonged video-EEG monitoring study should be performed to capture the events and confirm the diagnosis. Neuroimaging studies, computed tomography (CT), and preferably MRI should be performed to search for structural brain abnormalities. If adrenocorticotropic hormone (ACTH) or corticosteroids are to be used to treat the patient, these neuroimaging studies should be obtained before institution of such therapy because these agents are known to produce enlargement of the cerebrospinal fluid spaces. Such changes are very difficult to distinguish from preexisting cerebral atrophy. A variety of routine laboratory tests should be obtained, including complete blood count with differential, liver panel, renal panel, electrolytes and glucose, serum magnesium, calcium and phosphorous, and urinalysis prior to initiating treatment. If an associated condition is not identified after completion of these routine studies, a more thorough workup should begin. Metabolic studies including plasma ammonia, serum lactate and pyruvate, urine organic acids, serum biotinidase, and serum and urine amino acids should be performed. Chromosomal analysis should be obtained. Finally, cerebrospinal fluid should be evaluated for cell count, protein, glucose, lactate, pyruvate and amino acids, and bacterial and viral culture.

On the basis of these data, patients can be divided into two groups: cryptogenic or symptomatic. Cryptogenic patients are those with no abnormality on neurological examination, normal development before onset of spasms, no known etiological factor, and normal neuroimaging studies. About 20% of patients are currently classified as cryptogenic. The remaining 80% of patients who fail to meet one or more of these criteria are classified as symptomatic. Some investigators identify a third group of patients, designated idiopathic, which includes patients presumed to be of genetic origin. This information can be helpful in predicting long-term outcome, as cryptogenic patients have the best prognosis for normal developmental outcomes.

TREATMENT STRATEGY

Contrasting opinions have evolved over the decades, due to the many methodological shortcomings of published efficacy studies, as to the best treatment of infantile spasms. Recent critical reviews of this subject have concluded that ACTH is effective in the short-term treatment of infantile spasms and that vigabatrin may be effective in stopping the spasms in patients with tuberous sclerosis. However, these reviews found insufficient evidence to recommend any other treatment for this disorder, nor was there sufficient evidence to conclude that treatment resulting in control of spasms improves long-term outcome.

Similarly, the available data concerning the surgical treatment of infantile spasms patients with focal abnormalities on EEG, CT, MRI, or positron emission tomography (PET) does not allow definitive conclusions to be reached. However, focal cortical resection or hemispherectomy may be beneficial in a select group of infantile spasms patients with focal cortical abnormalities who have failed drug therapy.

Based on our own analysis of the data, many seemingly unrelated therapeutic modalities have shown some efficacy in the treatment of infantile spasms. The dosages and durations of treatment, side effects, formulations, proposed mechanisms of actions, and response characteristics of each of these agents can be found in our review of the topic (Frost and Hrachovy 2003). Most investigators believe that ACTH is the most effective agent; however, there is no convincing evidence that higher doses of ACTH are more effective than lower doses of the drug. Vigabatrin appears to be particularly effective in controlling the spasms in patients with tuberous sclerosis. Response to any form of therapy usually occurs within 1–2 weeks, and there are no factors (e.g., patient classification or treatment lag) that can definitely be used to predict response to therapy. On the basis of this analysis, we recommend the following systematic therapeutic approach (Figure 14.1): The primary goal is to stop the spasms and to improve the EEG as soon as possible, and to avoid prolonged treatment with any specific mode of therapy. If the patient fails to respond to one agent within the recommended time interval, it should be immediately stopped and a new agent tried. The specific therapeutic guidelines for each modality are shown in Table 14.1.

We believe that prolonged video-EEG monitoring is the best method to objectively assess treatment response. However, if such monitoring is not possible, the physician will have to rely on the results of routine EEGs and caregiver observations to determine response to therapy. If spasms are not seen during an intense observation period of at least five consecutive days, and if the repeat EEG, including a sleep recording, has improved, then it can be assumed that a response has occurred. If a relapse occurs following discontinuation of therapy, the agent that previously produced the response should be restarted.

LONG-TERM OUTCOME

There is no conclusive evidence that medical treatment of this disorder, even when associated with successful control of the spasms, alters the developmental/mental outcome. In our analysis of the long-term outcome in studies with at least 25 patients per study and an average duration of follow-up of 31 months (Frost and Hrachovy, 2003), we found that only 16% of patients in these studies had normal development at follow-up, 47% of patients continued to experience seizures at follow-up, and seizure rates were higher in symptomatic patients (54%) compared to cryptogenic patients (23%). The most common seizures observed were simple partial, tonic, and generalized tonic–clonic seizures. The Lennox–Gastaut syndrome developed in 17% of patients. Approximately 44% of patients had persisting neurological deficits, and 61% had abnormal EEGs. The average mortality rate was 12%, a rate that has slightly declined over the decades. The most important factor predictive of a normal outcome was classification into the cryptogenic category. It was seen that 51% of cryptogenic patients had normal development compared to 6% of symptomatic patients. The only other positive prognostic indicators were sustained control of spasms and absence of other seizure types.

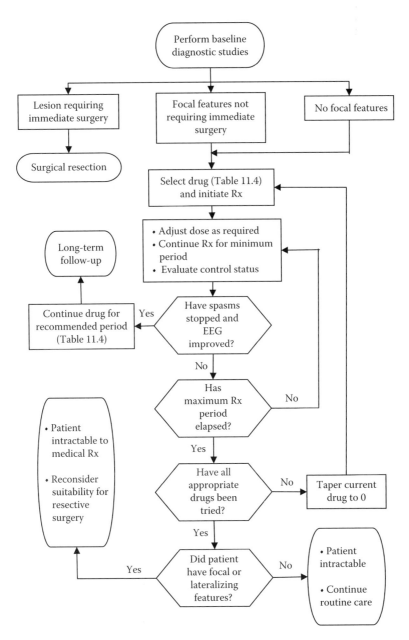

FIGURE 14.1 Flowchart summarizing recommended approach to the treatment of infantile spasms. (From Frost, JD, Jr, and Hrachovy, RA. [2003]. *Infantile Spasms: Diagnosis, Management, and Prognosis.* Kluwer Academic, New York. With permission.)

PATHOPHYSIOLOGY/NEUROBIOLOGY OF DISEASE

The pathophysiological mechanism underlying infantile spasms is not known.

TABLE 14.1

Therapeutic modalities with demonstrated efficacy in infantile spasms and suggested parameters for implementation

Therapy	Initial dose	Maximum maintenance dose	Minimum duration of therapy	Maximum duration of therapy if no response	Continue therapy if response occurs?
ACTH	20 u/day	30 u/day	2 weeks (plus 1 week taper)	6 weeks (plus 1 week taper)	No
Corticosteroid (prednisone)	2 mg/kg/day	2 mg/kg/day	2 weeks (plus 1 week taper)	6 weeks (plus 1 week taper)	No
Vigabatrin[a]	50 mg/kg/day	200 mg/kg/day	N/A	8 weeks	Yes[b]
Nitrazepam[a]	1 mg/kg/day	10 mg/kg/day	N/A	12 weeks	Yes[b]
Valproate	40 mg/kg/day	100 mg/kg/day	N/A	8 weeks	Yes[b]
Pyridoxine (vitamin B$_6$)	100 mg/day or 20 mg/kg/day	400 mg/day or 50 mg/kg/day	1 week	2 weeks	Yes[b]
Topiramate	12 mg/kg/day	24 mg/kg/day	N/A	8 weeks	Yes[b]
Zonisamide	3 mg/kg/day	13 mg/kg/day	N/A	6 weeks	Yes[b]
Immunoglobulin	100–400 mg/kg/day 1–5 days	400 mg/kg/day 5 days every 6 weeks	5 days	8 weeks	Yes, up to 6 months
TRH	0.05–0.5 mg/kg/day	1.0 mg/kg/day	2 weeks	4 weeks	No
Surgery	N/A	N/A	N/A	N/A	N/A

Note: N/A = not applicable to this form of therapy.

[a] These drugs are not approved for general use in the United States.

[b] An attempt at discontinuation is suggested after several months.

Source: Frost JD, Jr, and Hrachovy, RA. (2003). *Infantile Spasms: Diagnosis, Management, and Prognosis.* Kluwer Academic, New York. With permission.

For those interested in a thorough discussion of the proposed pathophysiological mechanisms underlying this disorder, it is suggested that the reader review our recently published paper on the topic (Frost and Hrachovy, 2005). A brief review of some of these hypotheses follow.

Early studies implicated the brainstem as the area generating the epileptic spasms and producing the hypsarrhythmic EEG pattern. Subsequently, it was hypothesized that the brainstem dysfunction causing infantile spasms was a result of an abnormal functional interaction between the brainstem and a focal or diffuse cortical abnormality. According to this hypothesis, the cortical abnormality exerts a noxious

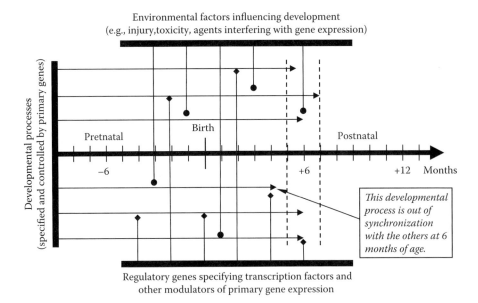

FIGURE 14.2 Developmental desynchronization model of infantile spasms pathogenesis showing schematically the interaction of developmental processes controlled by primary genes (e.g., neurogenesis, myelination, synaptogenesis, apoptosis, neurotransmitter systems [horizontal lines]) with regulatory gene effects (vertical lines from bottom) and environmental factors (vertical lines from top). Vertical dashed lines indicate hypothetical maximal extent of desynchronization consistent with normal function at 6 months. (From Frost, JD, Jr and Hrachovy, RA. [2005]. Pathogenesis of infantile spasms: A model based on developmental desynchronization. *J Clin Neurophysiol* 22[1], 22–36. With permission.)

influence over the brainstem from where the discharges spread caudally and rostrally to produce the spasms and the hypsarrhythmic EEG pattern. This hypothesis was based primarily on the results of PET scan studies and, subsequently, on the association of partial seizures with infantile spasms.

Other major hypotheses for the cause of infantile spasms include a defect in the immune system, dysfunction of neurotransmitter systems, and stress or injury during early infancy, resulting in the release of excess amounts of corticotrophin-releasing hormone (CRH).

Recently, we proposed a new model concerning the pathophysiology of this disorder, based on the concept of developmental desynchronization. According to this model, infantile spasms result from a particular temporal desynchronization of two or more developmental processes. As illustrated in Figure 14.2, the developmental desynchronization could be produced by (1) a mutation or inherited abnormality affecting the primary genes governing ontogenesis, (2) a mutation or inherited abnormality affecting genes specifying transcription factors or other genetic modulators, or (3) an external environmental factor affecting the maturational processes of brain tissues and/or neurochemical systems. Each mechanism (or combination of mechanisms) could manifest at different locations and at different points of development.

This model is supported by our recent animal work where neuronal blockade produced by infusion of tetrodotoxin into the cortex or hippocampus of neonatal rats produced interictal and ictal EEG changes and epileptic spasms similar to those seen in humans.

CLINICAL PEARLS

1. A clinical history of spasm-like events that occur in clusters, especially upon arousal from sleep, strongly suggests the diagnosis of infantile spasms.
2. Long-term video-EEG monitoring is often needed to confirm the diagnosis of infantile spasms.
3. Patients classified as cryptogenic have the best prognosis for normal developmental outcome.
4. Although ACTH and vigabatrin are recognized as the most effective therapeutic agents to treat infantile spasms, many other therapeutic modalities exist.
5. Response to any therapeutic regimen usually occurs within 1–2 weeks, that is, prolonged treatment with any agent should be avoided.

SUGGESTED READING

Chugani HT, Shewmon DA, Sankar R, Chen BC, Phelps ME. (1992). Infantile spasms II. Lenticular nuclei and brain stem activation on positron emission tomography. *Ann. Neurol.* 31, 212–219.

Frost JD, Hrachovy RA. (2003). *Infantile Spasms: Diagnosis, Management, and Prognosis.* Kluwer Academic, New York.

Frost JD, Hrachovy RA. (2005). Pathogenesis of infantile spasms: A model based on developmental desynchronization. *J. Clin. Neurophysiol.* 22, 22–36.

Hancock E, Osborne J. (2003). Treatment of infantile spasms. *The Cochrane Database Syst Rev. Issue 3* : CD001770.

Hrachovy RA, Frost JD. (1989). Infantile spasms: a disorder of the developing nervous system. In *Problems and Concepts in Developmental Neurophysiology,* Kellaway P and Noebels JL, Eds. 131–147. Baltimore: John Hopkins University Press.

Hrachovy RA, Frost JD. (2003). Infantile epileptic encephalopathy with hypsarrhythmia (Infantile Spasms/West Syndrome). *J. Clin. Neurophysiol.* 20, 408–425.

Hrachovy RA, Frost JD, Kellaway P. (1984). Hypsarrhythmia: Variations on the theme. *Epilepsia* 25, 317–325.

Mackay MT, Weiss SK, Adams-Webber T. et al. (2004). Practice Parameter: Medical Treatment of Infantile Spasms: Report of the American Academy of Neurology and the Child Neurology Society. *Neurology* 62, 1668–1681.

15 Gelastic Seizures

Yu-tze Ng, M.D., FRACP

CONTENTS

CASE PRESENTATION

A developmentally normal, 30-month-old boy began having gelastic (i.e., associated with mirth) seizures at the age of four months. His past history was significant only for a Nissen fundoplication, which may have been performed for presumptive gastroesophageal reflux disease (GERD) or, more likely, gelastic seizures mistaken for GERD. The seizures were stereotyped and characterized by sucking and laughing. Often, the patient would ask for a drink during the seizure and would drink ferociously if not restrained. At times, the patient would also become violent. The seizures were brief and averaged 30 seconds in duration (range 10–90 seconds) with only occasional, minimal postictal lethargy. He was subsequently diagnosed with a hypothalamic hamartoma (HH) on brain MRI scanning. Seizure frequency had been variable initially, but gradually evolved to an average of every 5 minutes, constituting "status gelasticus." The seizures would persist through sleep and awaken the patient throughout the night. The patient had previously failed therapy with phenobarbital, topiramate, and clonazepam. He was treated with levetiracetam, acetazolamide, and nocturnal high-dose lorazepam. None of the antiepileptic drugs (AEDs) significantly reduced seizure frequency. His neurological examination was otherwise normal. The patient was transferred to a tertiary center for emergent surgical treatment/resection of the HH lesion. Twenty-four hour scalp video-EEG recording was performed as well as a preoperative brain MRI scan that showed his HH (Figure 15.1A and B). Video-EEG recording confirmed an average of 10 gelastic seizures per hour as identified by parents. There were no ictal EEG patterns seen other than muscle and motion artifact. In addition, the patient's baseline and interictal recordings were

117

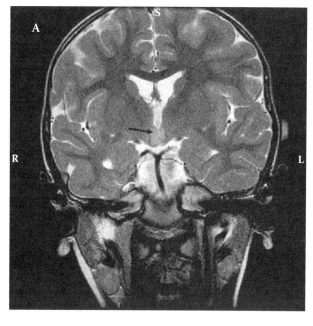

(A)

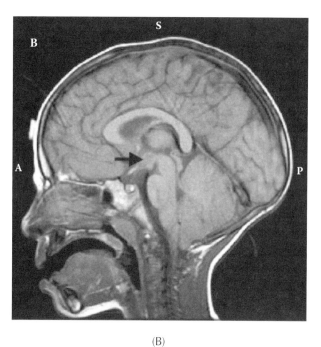

(B)

FIGURE 15.1   Preoperative brain MRI T2-weighted coronal (A) and T1-weighted sagittal (B) views of the hypothalamic hamartoma as shown by the arrows.

normal. The patient then underwent emergent transcallosal interforniceal resection of the HH. The surgery was complicated by a small right-sided thalamic infarct with resultant mild transient left hemiparesis that completely resolved within 2 days. Figures 15.2A and B show the postoperative brain MRI scan. The patient had three brief (less than 30 seconds) stereotypical seizures within the first week after surgery. He became seizure-free for 2 months before the gelastic seizures recurred, but at a much reduced seizure frequency (i.e., >90% reduction compared to his preoperative baseline). Neuropathological examination of the resected lesion demonstrated subependymal tissue composed of disorganized glial and neuronal elements consistent with an HH. After 19 months, endoscopic resection (via the lateral ventricle and through the foramen of Monro) of residual hypothalamic hamartoma tissue was performed for persistent gelastic seizures. The patient has now been seizure free for more than 12 months, and off antiepileptic medications. He is assessed to be developmentally normal with minor behavioral problems (hyperactivity with labile mood).

DIFFERENTIAL DIAGNOSIS

Gelastic (or laughing) seizures were first described by Daly and Mulder in 1957. They are characterized by bouts of laughter that may be either similar or, more commonly, distinct from the patients' usual laughter, and associated with a slight sensation or appearance of discomfort. A related seizure type may involve crying and/or facial contraction with an exaggerated grimace; these are referred to as dacrystic seizures. Affected patients may exhibit both forms of seizures, or seizures with mixed features of both types. Autonomic symptoms such as flushing, tachycardia, and altered respiration are often associated with these seizures. Most (but not all) of these seizures are simple partial in nature with preservation of awareness. They are usually brief (less than 30 seconds) without a postictal phase. Status gelasticus is the most severe form, defined as a prolonged cluster of gelastic seizures lasting longer than 20–30 minutes. Patients usually do not report a feeling of mirth. In its mildest form, patients have simply described an urge to laugh that can be self suppressed. Scalp EEG monitoring usually does not show any ictal correlate. Typically, the seizure diagnosis is missed or delayed for many years and is often misdiagnosed as a "happy baby," colic, or gastroesophageal reflux disease. Patients with gelastic seizures should undergo detailed neuroimaging with particular emphasis on the hypothalamus, including MRI brain scans with fine coronal sections through this region.

Most cases of gelastic seizures represent a symptomatic form of epilepsy. By far the commonest etiology of gelastic epilepsy is an HH. HHs are rare developmental malformations of the inferior hypothalamus and tuber cinereum. Other causes include rare structural lesions impinging upon the floor of the third ventricle, such as tubers of tuberous sclerosis, pituitary tumors, gliomas, meningiomas, and basilar artery aneurysms. Frontal and temporal lobe epilepsy rarely cause gelastic seizures.

In HH patients, gelastic seizures are almost always the first seizure manifestation, which in retrospect often begins shortly after birth. Many of these patients subsequently develop a refractory mixed epilepsy and epileptic encephalopathy. It

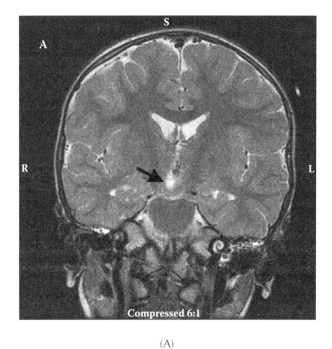

(A)

(B)

FIGURE 15.2   Postoperative T2-weighted brain MRI coronal (A) and sagittal (B) images of the resected hypothalamic hamartoma shown by the arrows. The postoperative drain tube that was subsequently removed is seen on the sagittal view.

is believed that the other evolving seizure types result from a secondary "epilepto-genesis" where other parts of the brain "learn" from the HH how to generate sei-zures. Mental retardation and behavioral problems—including rage attacks—are commonly seen. In addition, precocious puberty occurs in approximately half the HH patients. A subset of HH patients has a specific midline syndrome known as Pal-lister–Hall syndrome. It is a rare syndrome that can occur either spontaneously or be inherited in an autosomal dominant fashion through a mutation in the *GLI3* gene. It is associated with polydactly, midline defects, including dysmorphic facial features, hypothalamic hamartoma, and imperforate anus.

CLINICAL APPROACH

Although HH is relatively uncommon with a prevalence of about 1 in 100,000, it is almost certainly underdiagnosed by medical caregivers. The gelastic seizures may initially not be typical, and the parents may simply be aware that something is wrong or that they have a baby who "laughs too much." Equally common, the seizures may be more of a dycrastic, or crying, seizure, sometimes associated with strained, painful, and paroxysmal but stereotypical discomfort spells that may resemble gastroesophageal reflux disease or colic. Any patient with typical gelastic seizures should be presumed to have an HH and be evaluated with appropriate neuroimaging at an experienced/tertiary medical center.

Central precocious puberty (CPP), which affects many HH patients, is another very distinctive symptom that should alert one to the diagnosis of an HH. The asso-ciated refractory mixed epilepsy and epileptic encephalopathy are less specific but certainly part of the clinical picture. Although other seizure types often present later, that is not always the rule, and in fact, HHs are an uncommon but important cause of infantile spasms (initial presentation).

NEUROBIOLOGY/PATHOPHYSIOLOGY OF DISEASE

The expression of laughter appears to depend on two different neuronal pathways. One is an involuntary system that involves the deep gray matter structures includ-ing the amygdala, thalamic and subthalamic areas, and the dorsal tegmentum; the second and voluntary system originates in the premotor frontal opercular areas and leads through the motor cortex and pyramidal tract to the ventral brainstem. The laughter may result from a laughter-coordination center in the dorsal upper pons.

The pathophysiology of gelastic seizures (and secondary epileptogenesis) aris-ing from HH tissue is poorly understood. However, initial studies have revealed two distinct populations of neurons in surgically resected HH tissue. The first group consists of small γ-aminobutyric acid (GABA)-expressing neurons found principally in nodules that display spontaneous rhythmic firing. The second population is com-posed of large, quiescent, pyramidal-like neurons with more extensive dendritic and axonal arborization. It has been proposed that the small, spontaneously firing GAB-Aergic neurons might send inhibitory projections to, and drive the synchrony of, large output HH neurons. Alternatively, the majority of large HH neurons have been found to depolarize in response to $GABA_A$ receptor activation, and such an effect could lead to neuronal excitation.

TREATMENT AND LONG-TERM OUTCOME

Typically, in HH patients, gelastic seizures (and also other seizure types) are extremely refractory to antiepileptic drugs and other nonpharmacological therapies. Even as recently as the past decade, experts felt that neurosurgical resection could not be performed safely due to location, and even if it could, might not help the epilepsy. Both of these notions have now been dispelled, and relatively large series of patients have now been cured of their refractory symptomatic gelastic and mixed epilepsy.

For those patients who fail to respond to medications, surgical resection using a transcallosal, interforniceal approach has been shown to be efficacious and generally safe. More recently, many HH surgical resections have been performed using an endoscopic technique with a transventricular approach. Gamma knife surgery has also been used to treat several HH patients and is often advocated for smaller lesions, particularly in Europe.

The natural history of HHs is generally very poor with a progressive epileptic encephalopathy, severe mental retardation, and persistent refractory epilepsy. However, variability exists with milder (possibly even asymptomatic) cases never being diagnosed. Indeed, typically with the pedunculated form of HH, some patients may present solely with precocious puberty. More severely affected patients can be significantly improved with surgical therapy resulting in around half the patients becoming seizure free and nearly 90% with significant seizure reduction. In addition, many of these patients and their families report postsurgical behavioral and cognitive improvement. Precocious puberty should be evaluated and followed by a pediatric endocrinologist. Lupron® (leuprolide acetate), a gonadotropin-releasing hormone (GnRH) agonist, is effective in treating precocious puberty. CPP may resolve following resection of the HH.

CLINICAL PEARLS

1. Although uncommon, HHs are probably underdiagnosed, and caregivers should be aware of the usual (but not always) trademark-presenting gelastic or laughing seizures.
2. Other diagnostic clues include CPP, refractory mixed epilepsy, and an idiopathic psychiatric scenario including autistic features, mental retardation, and behavioral problems—in particular, rage attacks.
3. HHs may present as part of a midline syndrome, in particular Pallister–Hall syndrome, with typical clinical features of polydactly and midline defects, including dysmorphic facial features and imperforate anus.
4. Medical therapy with AEDs is unlikely to provide seizure freedom, and there should be early consideration for surgical treatment.

SUGGESTED READING

Daly D, Mulder D. (1957) Gelastic epilepsy. *Neurology* 7: 189–92.

Fenoglio KA, Wu J, Kim do Y et al. (2007) Hypothalamic hamartoma: basic mechanisms of intrinsic epileptogenesis. *Semin. Pediatr. Neurol.* 14(2): 51–59.

Freeman JL, Harvey AS, Rosenfeld JV, Wrennall JA, Bailey CA, Berkovic SF. (2003) Generalized epilepsy in hypothalamic hamartoma: evolution and postoperative resolution. *Neurology* 60: 762–767.

Kerrigan JF, Ng YT, Chung S, Rekate HL. (2005) The hypothalamic hamartoma: a model of subcortical epileptogenesis and encephalopathy. *Semin. Pediatr. Neurol.* 12(2): 119–131.

Kerrigan JF, Ng YT, Prenger E, Krishnamoorthy KS, Wang NC, Rekate HL. (2007) Hypothalamic hamartoma and infantile spasms. *Epilepsia* 48: 89–95.

Kim DY, Fenoglio KA, Simeone TA et al. (2008) GABA(A) receptor-mediated activation of L-type calcium channels induces neuronal excitation in surgically resected human hypothalamic hamartomas. *Epilepsia* 49(5): 861–871.

Ng YT, Rekate HL. (2006) Coining of a new term, "Status Gelasticus." *Epilepsia* 47: 661–662.

Ng YT, Rekate HL, Prenger EC et al. (2008) Endoscopic resection of hypothalamic hamartomas for refractory symptomatic epilepsy. *Neurology* 70(17): 1543–1548.

Sweetman LL, Ng YT, Kerrigan JF. (2007) Gelastic seizures misdiagnosed as gastro-esophageal reflux disease. *Clinic. Peds.* 46: 325–328.

Wild B, Rodden FA, Grodd W, Ruch W. (2003) Neuronal correlates of laughter and humor. *Brain* 126: 2121–2138.

16 Tuberous Sclerosis Complex

Aimee F. Luat, M.D. and Harry T. Chugani, M.D.

CONTENTS

CASE PRESENTATION

Our patient was 2½-years-old when she first presented to us for further evaluation and management of her intractable seizures. She was an Egyptian girl born full term after an uneventful pregnancy and vaginal delivery. At birth, several hypopigmented macules were noted on her skin. She began having seizures on the first day of life. Seizure semiology consisted of eye gaze to one side followed by generalized clonic activity. Neurologic investigations included lumbar puncture, metabolic studies, electroencephalography (EEG), and cranial magnetic resonance imaging (MRI). MRI showed multiple bilateral cortical tubers. Echocardiogram disclosed the presence of cardiac rhabdomyomas. With these clinical findings, a definite diagnosis of tuberous sclerosis complex (TSC) was made. Her seizures became controlled with phenobarbital for one and a half years and, thereafter, the medication was discontinued. However, the seizures recurred and persisted despite trials of valproic acid, clonazepam, phenobarbital, and lamotrigine. Subsequently, she developed epileptic spasms consisting of sudden and brief flexion of her neck, arms, and legs against her body. These episodes occurred in clusters, especially during drowsiness and upon arousal from sleep. Initially, she had only two or three spasms per cluster, but later on episodes increased up to 20 individual spasms in a cluster. She had globally delayed development, and at the age of 2½-years, she had no words and could not walk without support. There was no family history of tuberous sclerosis or epilepsy. On physical examination, she was microcephalic with head circumference of 46 cm (less than second percentile). Multiple hypopigmented macules were noted on her face and trunk. On neurological examination, she was awake and alert, but she spoke no words. Her neurologic examination was normal except for her inability to walk without support.

Upon presentation to us, cranial MRI was repeated, and it showed multiple cortical tubers in the left and right hemispheres (Figure 16.1). A calcified tuber was noted in the right inferior frontal gyrus (white arrow in figure). Multiple calcified subependymal nodules were also noted along the lateral ventricles. TSC DNA testing confirmed the presence of a *TSC2* gene mutation. She was started on vigabatrin at 50 mg/kg/day, and this was later increased to 80 mg/kg/day. Her seizures became fairly controlled with vigabatrin for a year and a half, such that she would have breakthrough seizures only when she was ill. However, her seizures subsequently increased in frequency despite increased dosage of vigabatrin and the addition of oxcarbazepine. Due to the medical intractability of her seizures, she was evaluated for epilepsy surgery. Video-EEG captured focal seizures consisting of behavioral arrest, staring, and unresponsiveness, followed by clusters of epileptic spasms. Ictal EEG showed seizure onset from the right frontal-temporal region. Subclinical seizures coming from the right frontal region were also captured. Interictally, multifocal spike-and-wave activities were noted. A 2-Deoxy-2-[18F] fluoro-D-glucose (FDG) positron emission tomography (PET) scan showed multiple areas of glucose hypometabolism in both the left and right hemispheres (Figure 16.2A, black arrows); $\alpha[^{11}C]$ methyl-L-tryptophan (AMT) PET scan showed intense uptake in a right frontal tuber (Figure 16.2B). The patient underwent a two-stage epilepsy surgery with extraoperative electrocorticography (ECOG). Numerous clinical as well as electrographic seizures of right frontal onset were captured. She underwent right frontal lobectomy guided by the findings of ECOG and the AMT PET. Pathology showed multiple areas of dysplastic cortex with loss of normal laminar architecture. Increased fibrillarity of the neurophil was also noted. Dysplastic cells including balloon cells and cytomegalic neurons were noted (Figure 16.3). On 12 months' follow-up, the child was seizure free. Progress in her development has also been noted. Follow-up EEG showed rare polyspike-and-wave activities from the right temporal region. No electrographic seizures were noted.

DIFFERENTIAL DIAGNOSIS

The causes of intractable epilepsy in children are heterogeneous. In the newborn period, intractable epilepsy is rarely idiopathic. Hence, extensive neurologic investigations should be performed in order to establish and, possibly, to treat the cause. In neonates, a broad range of systemic and central nervous system disorders can increase the risk for seizures, including hypoxic-ischemic encephalopathy, intracranial hemorrhage, and cerebral malformations. TSC has rarely been reported as a

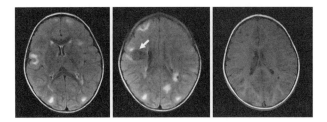

FIGURE 16.1 Fluid attenuated inversion recovery (FLAIR) pulse sequence MRI showed multiple and extensive areas of high signal intensity in both the left and right cerebral hemispheres, consistent with multiple cortical and subcortical tubers. A calcified tuber located in the right inferior frontal gyrus (arrow) can be noted showing low signal intensity on FLAIR.

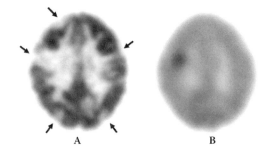

FIGURE 16.2 (A) 2-Deoxy-2-[18F] fluoro-D-glucose (FDG) positron emission tomography (PET) scan showed multiple areas of glucose hypometabolism (black arrows) in both the left and right hemispheres; (B) α[^{11}C] methyl-L-tryptophan (AMT) PET scan showed intense uptake in a right frontal tuber.

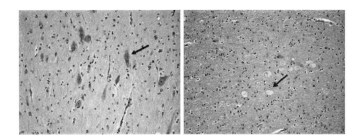

FIGURE 16.3 Histopathology of the resected right frontal cortex using hematoxylin and eosin stain (H and E) showing the presence of giant cells (left arrow) and balloon cells (right arrow).

cause of neonatal seizures. The frequent absence of the traditional stigmata of TSC in neonates may account for the underdiagnosis of TSC in this age group. Therefore, a high index of suspicion for TSC in every neonate who presents with idiopathic intractable seizures is necessary. In our case, the diagnosis of TSC was straightforward with its typical clinical findings, namely, the presence of hypopigmented macules, cardiac rhabdomyomas, and intractable seizures. It should be noted that

TABLE 16.1

Revised clinical diagnostic criteria for tuberous sclerosis complex (TSC)

Major features	Minor features
Facial angiofibromas or forehead plaques	Multiple, randomly distributed pits in dental enamel
Nontraumatic ungual or periungual fibroma	Hamartomatous rectal polyps
Hypomelanotic macules (three or more)	Bone cysts
Shagreen patch (connective tissue nevus)	Cerebral white matter radial migration lines
Cortical tuber	Gingival fibromas
Subependymal nodule	Nonrenal hamartoma
Subependymal giant-cell astrocytoma	Retinal achromic patch
Cardiac rhabdomyoma, single or multiple	"Confetti" skin lesions
Pulmonary lymphangiomyomatosis and/or renal angiomyolipomas	Multiple renal cysts
Multiple retinal nodular hamartomas	

Notes:

Definite TSC: 2 major or 1 major plus 2 minor features

Probable TSC: 1 major plus 1 minor feature

Possible TSC: 1 major or 2 minor features

the onset of either partial seizures or epileptic spasms in infants with hypomelanotic macules, as in our case, strongly suggest the diagnosis of TSC.

DIAGNOSTIC APPROACH

TSC is characterized by the development of hamartomas in multiple organs of the body, including the skin, brain, kidneys, heart, and the eyes. Its clinical diagnosis has been revised into major and minor features, and this classification provides the most current approach to the accurate clinical diagnosis of TSC (Table 16.1).

The diagnosis of TSC in newborns can be difficult because, in most infants, the skin and visceral lesions may not be apparent. The use of a Wood's (ultraviolet) lamp may allow for the detection of small or subtle skin lesions. In our patient's case, the diagnosis was clear because the clinical features of TSC were readily apparent. She presented with the typical neurocutaneous stigmata of TSC, as well as the other major clinical features, which included the presence of cardiac rhabdomyoma and cortical tubers. Neuroimaging, preferably with MRI, echocardiography, and renal ultrasonography should be performed to confirm the diagnosis. Genetic testing for the causative genes of tuberous sclerosis, TSC1 and TSC2, are commercially available and should be considered for genetic counseling or in ambiguous cases.

TREATMENT STRATEGY

Epilepsy in TSC is often resistant to antiepileptic drugs (AED) and may have a negative impact on the child's neurocognitive development; hence, there is some urgency in achieving seizure control. Basic science studies have found enhanced expression of *N*-methyl-D-aspartate (NMDA) receptors and reduced expression of gamma-aminobutyric acid (GABA) receptors in both dysplastic neurons and giant cells. The impairment of GABAergic neurotransmission has been hypothesized as the basis behind the efficacy of vigabatrin in the treatment of epilepsy in TSC. Vigabatrin is a structural analog of GABA that produces its antiepileptic effect by irreversibly inhibiting GABA transaminase (GABA-T), the degradative enzyme for GABA, thus resulting in a significant rise in the brain and cerebrospinal fluid concentration of GABA and increased inhibition. Vigabatrin is able to completely stop infantile spasms in 95% of infants with TSC. It has been recommended as first-line therapy of infantile spasms due to TSC; however, concerns for ophthalmologic toxicity with prolonged use have limited its scope outside this indication.

Other newer drugs that have been utilized in the treatment of epilepsy in TSC include topiramate, lamotrigine, levetiracetam, and oxcarbazepine, and these medications appeared to be well tolerated and effective in small subgroups of patients with TSC. Despite the availability of vigabatrin and the new antiepileptic medications, epilepsy in TSC becomes medically intractable in many individuals. In such cases, resective epilepsy surgery may provide a good therapeutic option, especially if a single tuber is acting as the epileptogenic focus, as exemplified in our case.

Epilepsy surgery in TSC can be a challenge because the suspected epileptogenic tuber can be difficult to identify amidst the multiple bilateral lesions. Conventional MRI and EEG often show multifocal abnormalities. Likewise, FDG PET scan also show multifocal areas of hypometabolism without specifically indicating the epileptogenic region. AMT ($\alpha[^{11}C]$ methyl-L-tryptophan) is an analog of tryptophan, and AMT PET can be used to measure brain serotonin synthesis capacity noninvasively in humans. The use of AMT PET scanning has proven to be a useful tool in the identification of epileptogenic tubers and has improved the epilepsy surgery outcome in TSC. In our case, AMT PET showed intense uptake concordant with the ictal EEG onset zone, thereby strengthening the confidence of the localization of the potential epileptogenic zone. The precise localization of the epileptogenic zone was confirmed by the ECOG findings as well as by cessation of the patient's seizures after surgical resection of the right frontal lobe.

The ketogenic diet and vagus nerve stimulation have been shown to be effective and well tolerated treatments for refractory epilepsy. These alternative treatments should be strongly considered in those patients who are not appropriate surgical candidates.

PATHOPHYSIOLOGY/NEUROBIOLOGY OF THE DISEASE

TSC has an estimated birth incidence of about 1 in 10,000. It has an autosomal dominant pattern of inheritance, but 80% of TSC patients have no family history of the disease and, therefore, may represent a new mutation. Among the *familial* cases, about half are linked to TSC1 locus in chromosome 9q34 and half to TSC2 locus in chromosome 16p13.3. Among the *sporadic* cases, over 70% are due to TSC2 mutation.

Hamartin, a 130-kDa protein, is the product of *TSC1* gene. It inhibits tumor formation by regulating cell adhesion through its interaction with ezrin–radixin–moesin (ERM), family members of proteins that link the plasma membrane to the actin cytoskeleton, and by activation of the small guanosine triphosphate (GTP)-binding protein Rho. It has been proposed that loss of adhesion to the extracellular matrix due to loss of hamartin initiates the development of hamartomas in TSC.

On the other hand, *tuberin*, a 200-kDa protein, is a product of *TSC2* gene and is a GTPase-activating protein (GAP) for small GTP-binding protein Rap1. Rap 1 has been found to induce DNA synthesis, suggesting that it has a positive role in cellular growth regulation. Tuberin also has GTPase activity towards Rab5 and negatively regulates endocytosis. The loss of tuberin activity could interfere with the docking, fusion, and processing of the Rab5-GTP-associated early endosomes, and this could lead to missorting of internalized growth factor receptors or other signal-mediated membrane-bound molecules that would otherwise undergo lysosomal degradation.

Hamartin and tuberin interact with each other to form a cytoplasmic protein complex that inhibits the mammalian target of rapamycin (mTOR), which is the key protein and central regulator of cell growth. Therefore, mutations in the *TSC1* or *TSC2* genes that interfere with the assembly of functional tuberin–hamartin complexes cause unregulated activation of mTOR and lead to abnormal and dysregulated cell growth.

The tumorigenesis in several TSC lesions can also be explained by Knudson's two-hit hypothesis, in which a germline mutation in one allele of the *TSC1* or *TSC2* gene is followed by a second somatic mutation in the other allele, leading to cell growth derangement and hamartoma formation. This has been demonstrated in the development of subependymal giant cell astrocytomas (SEGA) and kidney angiomyolipoma.

EPILEPSY IN TUBEROUS SCLEROSIS COMPLEX

Epilepsy is the most common neurological feature of TSC occurring in 80% to over 90% of cases, often commencing in the first year of life. Infantile spasms and partial seizures are the most common seizure types. TSC is an important cause of infantile spasms, accounting for between 10 and 25% of cases. In our patient's case, she initially had partial seizures with secondary generalization, followed by the development of spasms with coexistent partial seizures.

Infantile spasms in TSC have distinctive clinical and EEG features. Each episode is usually associated with focal or lateralizing features such as tonic eye deviation, head turning, or nystagmus. Infants with epileptic spasms due to TSC exhibit a particular awake interictal EEG characterized by multifocal asynchronous spike discharges and irregular slow activity that increases and becomes generalized during non–rapid eye movement (non-REM) sleep. Hypsarrhythmia often appears later in the course. The ictal EEG in TSC may start with a focal discharge of spikes and polyspikes in the region of the epileptogenic tuber, followed by generalized irregular slow-wave and background attenuation. Our patient's ictal EEG demonstrated the phenomenon of focal seizures and epileptic spasms as a single ictal event, supporting the notion that cortical "triggering" mechanisms may be the underlying basis in the pathogenesis of epileptic spasms in certain groups of children.

Brain lesions in TSC responsible for epilepsy include cortical tubers and other associated cerebral/cortical malformations such as microdysgenesis and gray-matter heterotopias. Cortical tubers, the hallmark lesions in TSC, are considered to be errors of neuronal proliferation, differentiation, and migration, and have been implicated in the epileptogenesis in TSC.

Immunohistochemical studies have demonstrated that tubers express a variety of neuronal markers, neurotransmitter receptors, neuropeptides, and calcium-binding proteins, suggesting that they play a role in the pathogenesis of epileptic seizures. In addition, the expression of multidrug-resistance gene (MDR-1) and multidrug-resistance-associated protein-1 (MRP-1) in epileptogenic tubers has also been shown, suggesting that these factors may also contribute to the refractoriness of epilepsy in TSC to antiepileptic medications.

Because excessive astrocytosis has also been noted in cortical tubers, its role in epileptogenesis in TSC has been investigated in TSC transgenic animal models. Astrocytic proliferation and progressive epilepsy have been demonstrated in astrocyte-specific TSC1 conditional knockout mice (TSC1 cKO mice). The possible underlying mechanisms of seizures in this transgenic animal model include the altered glial glutamate transport secondary to decreased expression of astrocyte-specific glutamate transporters GLT-1 and GLAST, and the impairment of extracellular potassium uptake by the astrocytes through the astrocyte inward rectifier potassium (Kir) channels leading to neuronal hyperexcitability and epileptogenesis.

OUTCOME

Patients with TSC have a high prevalence of cognitive and behavioral difficulties, including autism. Infantile spasms and early intractable epilepsy may increase this risk, and side effects of polytherapy with antiepileptic medications may play a role. Many patients may respond to medical management with antiepileptic medications; however, up to 40% of patients with early onset seizures may prove medically refractory. Epilepsy surgery may render more than 50% of appropriately selected patients seizure free. A recent meta-analysis found that tonic seizures and moderate or severe intellectual disability were significant risk factors for seizure recurrence following surgery. For those that are not appropriate candidates, ketogenic diet or vagus nerve stimulation should be considered.

CLINICAL PEARLS

1. Tuberous sclerosis should be strongly considered in infants and children with seizures, particularly infantile spasm.
2. Vigabatrin is considered first-line therapy for infantile spasms associated with TSC.
3. In a subgroup of TSC patients whose seizures remain medically intractable, epilepsy surgery can be an appropriate treatment option, provided that a single epileptogenic focus can be demonstrated.

SUGGESTED READING

Cheadle JP, Reeve MP, Sampson JR, Kwiatkowski DJ (2000). Molecular genetic advances in tuberous sclerosis. *Hum. Genet.* 107, 97–114.

Curatolo P, Verdecchia M, Bombardieri R (2001). Vigabatrin for tuberous sclerosis complex. *Brain Dev.* 23, 649–53.

Curatolo P, Bombardieri R, Verdecchia M, Seri S (2005). Intractable seizures in tuberous sclerosis complex: from molecular pathogenesis to the rationale for treatment. *J. Child. Neurol.* 20, 318–325.

Curatolo P, Bombardieri R, Cerminara C (2006). Current management for epilepsy in tuberous sclerosis complex. *Curr. Opin. Neurol.* 19, 119–123.

Kagawa K, Chugani DC, Asano E et al. (2005). Epilepsy surgery outcome in children with tuberous sclerosis complex evaluated with alpha- [11C]methyl-L-tryptophan positron emission tomography (PET). *J. Child. Neurol.* 20, 429–38.

Lamb RF, Roy C, Diefenbach TJ et al. (2000). The TSC1 tumour suppressor hamartin regulates cell adhesion through ERM proteins and the GTPase Rho. *Nat. Cell. Biol.* 2, 281–287.

Lazarowski A, Lubieniecki F, Camarero S et al. (2004). Multidrug resistance proteins in tuberous sclerosis and refractory epilepsy. *Pediatr. Neurol.* 30, 102–6.

Roach ES, Gomez MR, Northrup H. (1998). Tuberous sclerosis complex consensus conference: revised clinical diagnostic criteria. *J. Child. Neurol.* 13, 624–628.

Tee AR, Fingar DC, Manning BD, Kwiatkowski DJ, Cantley LC, Blenis J (2002). Tuberous sclerosis complex- 1 and -2 gene products function together to inhibit mammalian target of rapamycin (mTOR)- mediated downstream signaling. *Proc. Natl. Acad. Sci. U S A* 99, 13571–13576.

White R, Hua Y, Scheithauer B, Lynch DR, Henske EP, Crino PB (2001). Selective alterations in glutamate and GABA receptor subunit mRNA expression in dysplastic neurons and giant cells of cortical tubers. *Ann. Neurol.* 49, 67–78.

Wienecke R, König A, DeClue JE (1995). Identification of tuberin, the tuberous sclerosis-2 product. Tuberin possesses specific Rap1GAP activity. *J. Biol. Chem.* 270, 16409–16414.

Wolf HK, Birkholz T, Wellmer J, Blumcke I, Pietsch T, Wiestler OD (1995). Neurochemical profile of glioneuronal lesions from patients with pharmacoresistant focal epilepsies. *J. Neuropathol. Exp. Neurol.* 54, 689–697.

Uhlmann EJ, Wong M, Baldwin RL et al. (2002). Astrocyte-specific TSC1 conditional knockout mice exhibit abnormal neuronal organization and seizures. *Ann. Neurol.* 52, 285–296.

17 Herpes Simplex Encephalitis

Dave F. Clarke, M.D.

CONTENTS

CASE PRESENTATION

The patient is an 11-year-old ambidextrous female with symptomatic intractable epilepsy who was referred for evaluation and treatment. At three years of age, she developed a fever, became confused, and within 2–3 hours began having persistent right hemibody myoclonic/clonic activity. She was admitted to an intensive care unit, and multiple antiepileptic drugs were required to fully control her seizures. Cerebrospinal fluid studies were positive for herpes simplex virus (HSV). Her initial EEG showed bilateral, independent, temporal, periodic epileptiform discharges (see Figure 17.1), and her initial neuroimaging studies revealed bilateral temporal edema. Though she was aggressively treated with acyclovir, the encephalitis caused significant impairment in (primarily short-term) memory, receptive and expressive language function (less than 15 words spoken intelligibly), and intractable epilepsy. However, she remembers individuals' names and several events prior to her neurological insult. Her present seizure semiology consists of predominantly generalized myoclonic and tonic events, with a frequency of 10–25 seizures per week. Prior medications include topiramate, phenobarbital, valproic acid (which caused hyperalbuminemia), oxcarbazepine, and phenytoin. A vagus nerve stimulator (VNS) was placed at the age of 9 years, but this was also unsuccessful in controlling her seizures. Her present antiepileptic medications include lamotrigine, felbamate, and zonisamide. A recent video-EEG study revealed interictal generalized and independent left and right

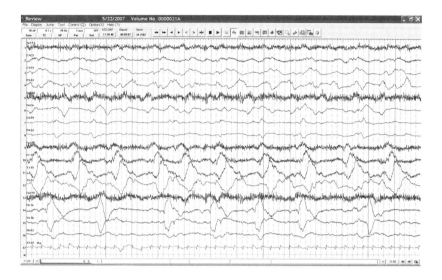

FIGURE 17.1   This is a 2-year-old, 9 days after clinical onset of herpes simplex encephalitis. EEG depicts right temporal periodic sharp waves with contralateral periodic slowing, which is maximal in the left temporal region. This is similar to the EEG described in the case illustration.

temporal epileptiform discharges (see Figure 17.2A). Two of ten seizures witnessed during monitoring were focal in onset, and these rapidly second-arily generalized. She also had generalized myoclonic/tonic events (see Figure 17.2B) that resembled flexor spasms (with elevation and flexion of both upper and lower extremities), and generalized myoclonic–astatic events during which she would fall forward. A brain MRI study revealed left temporal lobe encephalomalacia, as well as periventricular and subcortical white-matter signal hyperintensities; there was also widening of the sylvian fissure, and deepening and widening of sulci extending into the posterior portion of the left temporal lobe (see Figure 17.3). Neuropsychological testing revealed a func-tional level comparable to a 2- to 3-year-old child, and significant short-term memory deficits. After reviewing the results, a complete corpus callosotomy was recommended. At the 6-month follow-up visit, her mother described five brief (<1 minute) seizures, three of which involved right hemibody and two with left arm and face involvement. She has not had any generalized events nor has she fallen or injured herself since surgery.

DIFFERENTIAL DIAGNOSIS

Viral encephalitis is one of the most common causes of symptomatic status epi-lepticus. HSV, affecting about 1 or 2 cases per 500,000 per year, is the most com-mon cause of encephalitis in the United States. HSV acquired congenitally or in the

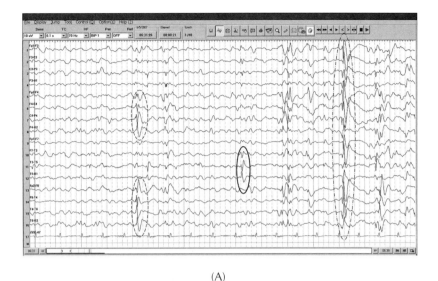

(A)

(B)

FIGURE 17.2 (A) Inter-ictal discharges (bipolar, anterior–posterior montage): (1) Right hemispheric discharge, maximal negativity in the temporal parietal region (initial dashed circles); (2) Left posterior temporal inter-ictal discharge (solid circle); (3) Generalized discharge followed by attenuation with no clinical correlate. (B) Myoclonic/tonic seizure (bipolar, anterior–posterior montage). Generalized polyspike and wave followed by low-amplitude faster-frequency activity with overriding myogenic artifact. This is interrupted by episodic right posterior quadrant discharges. Clinically, the patient had a generalized myoclonic jerk followed by tonic stiffening.

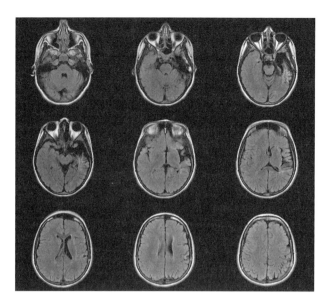

FIGURE 17.3 Axial FLAIR images depicting left temporal lobe encephalomalacia, and periventricular and subcortical white-matter signal hyperintensity primarily involving the left hemisphere. There was also widening of the sylvian fissure, and deepening and widening of sulci extending into the posterior portion of the left temporal lobe. An increase in the signal of the right medial temporal structures was also seen.

neonatal period (often HSV type 2) is a diffuse process with a different clinical presentation and course than that acquired during childhood (often HSV type 1). Though childhood-acquired HSV may have a wide spectrum of clinical presentations, it has a predilection for the limbic system and the temporal lobes. The case discussed represents a child who acquired herpes simplex encephalitis (HSE) at 3 years, which caused symptomatic intractable epilepsy and significant neurocognitive/neurodevelopmental deficits.

Fever and confusion should immediately alert the physician to the possibility of central nervous system (CNS) involvement. Partial seizures, though sometimes seen in other causes of encephalitis, are often a presenting symptom in herpes encephalitis. In a patient presenting with a seizure, meningitis, and other infectious processes, focal malignant or benign lesions, or a fever or illness that may have lowered the seizure threshold, have to be ruled out. Other symptoms suggestive of HSV include headache (irritability in younger children), unusual behavior, lethargy, vomiting, and other neurological symptoms such as cranial nerve findings and localized deficits. Symptoms are more nonspecific in very young children who may present with decreased activity or irritability and inconsolable crying.

DIAGNOSTIC APPROACH

A lumbar puncture is required in anyone in whom a CNS infection is suspected. Lymphocytic pleocytosis, elevated serum protein, and normal blood glucose are often seen in viral encephalitides, but normal values do not rule out the condition, and a mild decrease in glucose or a neutrophilic pleocytosis may be seen in the early stages. Hemorrhagic cerebrospinal fluid (CSF) is a sensitive indicator but is not specific to HSE. Polymerase chain reaction (PCR) of the CSF for herpes virus DNA, a test with both sensitivity and specificity above 90%, has been a significant diagnostic advancement. It is less invasive than the prior gold standard—a brain biopsy—and should therefore be carried out in anyone in whom encephalitis is suspected. False negatives may occur very early in the course of the disease; therefore, if the index of suspicion is high, a lumbar puncture should be repeated even after acyclovir has been started.

In HSE, the EEG is abnormal in over 90% of cases. Early changes in HSE consist of unilateral or bilateral, independent focal, or lateralized slowing that is maximal over the temporal and/or frontal lobes involved. This finding is due to the virus' predilection for infecting structures involved in the limbic circuitry (mesial temporal structures, isthmus, insula, orbitofrontal cortex, and cingulate gyrus). These EEG changes are followed by intermittent unilateral or bilateral sharp or slow-wave complexes, preceding periodic lateralized epileptiform discharges (PLEDs), or independent or symmetrical biperiodic lateralized epileptiform discharges (BiPLEDs) in the temporal regions. The discharges occur every 1 to 3 seconds. PLEDs are not specific for herpes but reflect acute destruction of the cerebral cortex, whether it is from a lesion or an acute infarct. The periodicity is usually seen 2 days to 1 week after onset, but may be seen later, and gradually disappears as the patient improves and the disease resolves. If the disease process persists, the complexes become broad, more prolonged suppression is seen after each burst and, in the final stages, there is more diffuse involvement preceding electrocerebral silence.

Hemorrhage seen on computer tomography (CT), primarily in the temporal lobes, is highly suggestive of HSE but is rarely seen in the early stages of the disease. The CT may be normal early in the course of the illness or may show hypodensities in the temporal lobes with mild mass effect. Patchy enhancement of gyri may also be seen when contrast is used. MRI is more sensitive and specific in the diagnosis of HSE than CT. The MRI initially reveals gyral edema on T1-weighted images and increased signal on T2 and FLAIR (fluid-attenuated inversion recovery) images in the temporal lobes, orbitofrontal cortex, and cigulate +/– the insula cortex. Petechial hemorrhages may be seen with MRI in the later stages of the disease, but, as with CT, are rarely seen in the early stages. With MRI contrast enhancement, these hemorrhages, usually absent in the early stages, become apparent as the disease progresses. The other limbic structures described become involved later. As the disease resolves, the long-term neuroradiological sequelae become apparent with destruction, encephalomalacia, and/or atrophy of portions of the temporal lobes and orbitofrontal lobes, primarily, as was seen in the patient described.

TREATMENT STRATEGY

Hospitalization is a necessity in children in whom encephalitis is suspected. In patients in whom HSE is a part of the differential diagnosis, acyclovir should be started immediately. Treatment is required for 14–21 days. Acyclovir is not without side effects: it is potentially nephrotoxic and may cause neutropenia. The dose may need to be adjusted in patients with impaired renal function, and frequent blood draws are required.

Seizures are a more common presenting symptom in patients with poor outcome and should be treated aggressively. Clinical seizures may be a treatment-and-diagnostic dilemma. PLEDs often represent the underlying process and have been described in rare cases of "burnt-out" status epilepticus. In a patient with prolonged status and prior documented subclinical seizures, a therapeutic trial is sometimes carried out. Benzodiazepines, phenytoin, and phenobarbital are the traditional antiepileptic drugs used to treat prolonged seizures. These agents—primarily the benzodiazepines and phenobarbital—may cause respiratory suppression and excessive drowsiness, thereby compromising the ability of the examiner to adequately determine neurological function. Whenever possible, less-sedating agents may be initiated if continued treatment is necessary. In cases where oral medications cannot be given, newer intravenous agents such as IV valproic acid or IV levetiracetam are available for use in most centers. In patients with persistent seizures, using appropriate antiepileptic agents for the long term is dependent on not only seizure type but also potential comorbidities. Treating mood and behavior, obesity, or weight loss in poorly functioning and/or immobile patients, or bone loss with inducing agents, etc., may influence the physician's choice of antiepileptic drug. In medically refractory patients, if only focal temporal lobe seizures are seen, temporal lobectomy is an option. In cases where there is bilateral involvement, palliative procedures such as the vagus nerve stimulator and, though less frequently used, corpus callosotomy to prevent atonic and tonic events may be offered.

LONG-TERM OUTCOME

Studies suggest that abnormal EEGs are seen in over 90% of patients, whereas an abnormal MRI was seen in about 85%. Significant neurologic sequelae are seen in over 60% of patients, including persistent seizures in nearly half of patients. Abnormal neuroimaging or abnormal EEG findings were more prevalent in patients with poor outcome. Delayed initiation of acyclovir was another predictor of poor outcome. As stated previously, limbic structures, including the mesial temporal structures, are often the neuroanatomical regions of concern; therefore, memory and other neuropsychological functions related to limbic circuitry, such as emotion, may be impaired. There are also varying degrees of neurocognitive impairment. In patients with symptomatic epilepsy, the seizure semiology is determined by the regions maximally involved and by seizure spread. Clinical and electroencephalographic temporal lobe onset seizures are most often seen. Seizures may, however, evolve from one or both hemispheres independently or may present as a more generalized picture, though MRI findings may be focal or multifocal, as seen in the patient described.

PATHOPHYSIOLOGY/NEUROBIOLOGY

Of the eight identified human herpes virus family types, HSV 1, often associated with orofacial infections, is most often identified in patients over 6 months of age with HSE. Perinatal or congenital HSV infections are often caused by HSV 2. HSV 1 must come in contact with a mucosal surface or broken skin in order to infect the individual; hence, contact with a person excreting HSV is required. Virions are transported by retrograde flow along axons from the entry point to the nuclei of a limited number of sensory neurons. Viral replication occurs at the site of infection and the dorsal root ganglion.

Distinctive pathological features of HSE include severe inflammation, congestion, hemorrhagic necrosis, and damage or destruction to both gray and white matter, primarily affecting the temporal/medial temporal and orbitofrontal cortex, though other regions may be involved. Approximately a third of cases of HSE are acquired by primary infection, and two-thirds occur after a period of latency. The olfactory tract and its close association with the limbic system make it a feasible pathway for HSV to gain access to the central nervous system, whereas the trigeminal nerve provides another possible route. Viral route of brain access, viral predilection for the temporal lobe and limbic structures, and the cause for viral reactivation are, however, poorly understood in humans.

CLINICAL PEARLS

1. Herpes simplex virus (HSV), affecting 1 or 2 cases per 500,000 per year, is the most common cause of encephalitis in the United States.
2. Though childhood-acquired HSV may have a wide spectrum of clinical presentations, it has a predilection for the limbic system and the temporal lobes.
3. In a patient presenting with a seizure, meningitis and other infectious processes, focal malignant or benign lesions, or a fever or illness that may have lowered the seizure threshold have to be ruled out.
4. In HSE, the EEG is abnormal in over 90% of cases.
5. Periodic lateralized epileptiform discharges, though not specific for HSE, is the most frequent electroencephalographic finding described in this condition.
6. Abnormal neuroimaging, abnormal EEG, and delayed initiation of antiviral therapy (acyclovir) are predictors of poor outcome.

SUGGESTED READING

Demaerel P, Wilms G, Robberecht W. et al. (1992) MRI of herpes simplex encephalitis. *Neuroradiology* 34, 490–493.

Elbers JM, Bitnun A, Richardson SE. et al. (2007). A 12-year prospective study of childhood herpes simplex encephalitis: is there a broader spectrum of disease? *Pediatrics* 119(2), e399–407.

Hsieh WB, Chiu NC, Hu KC, Ho CS, Huang FY. (2007). Outcome of herpes simplex encephalitis in children. *J. Microbiol. Immunol. Infect.* 40(1), 34–38.

Lai CW, Gragasin ME. (1988). Electroencephalography in herpes encephalitis. *J. Clin. Neurophysiol.* 5, 87–103.

Romero JR, Kimberlin DW. (2003). Molecular diagnosis of viral infections of the central nervous system. *Clin. Lab. Med.* 23(4), 843–865.

Weil AA, Glaser CA, Amad Z, Forghani B. (2002). Patients with suspected herpes simplex encephalitis: rethinking an initial negative polymerase chain reaction result. *Clin. Infect. Dis.* 34, 1154–1157.

Whitley RJ, Kimberlin DW. (2005). Herpes simplex encephalitis: children and adolescents. *Semin. Pediatr. Infect. Dis.* 16, 17–23.

18 Refractory Status Epilepticus

James J. Riviello, Jr., M.D.

CONTENTS

CASE PRESENTATION

A 10-year-old boy with a several-day history of fever, malaise, diarrhea, and emesis developed convulsive status epilepticus (CSE) requiring diazepam, phenobarbital, and phenytoin. Cranial computed tomography (CT) was unremarkable, and cerebrospinal fluid (CSF) contained six white blood cells/mm^3 (95% lymphocytes, 5% monocytes), with normal sugar and protein. Electroencephalogram (EEG) revealed diffuse slowing with occasional temporal spikes. He continued to have frequent seizures described as sudden staring episodes with head and eye deviation to the left, associated with cyanosis. He was intubated and transferred to the ICU where he was treated with a pentobarbital infusion. Pentobarbital was weaned 2 days later, but he then had a recurrence of his seizure. Sodium thiopental was given, followed by midazolam. When the seizures persisted, high-dose phenobarbital was used.

DIFFERENTIAL DIAGNOSIS/DIAGNOSTIC APPROACH

This case involves refractory status epilepticus (RSE), which occurs when seizure activity persists despite adequate therapy. In children, cases of RSE not responding to first-line therapy usually occur secondary to an acute symptomatic SE or related to an underlying progressive neurological disorder. Psychogenic seizures (or nonepileptic seizures) must also be considered when seizures persist despite treatment, but are very unusual in the younger child. Unusual motor movements, an "on/off"

survivors, no child with acute symptomatic RSE returned to baseline neurologic status, and all developed severe, refractory epilepsy. This could be in part related to the development of mesial temporal sclerosis after SE, which is currently being studied. Therefore, the outcome is definitely related to the etiology.

There are ethical considerations regarding use of prolonged HDST. The question of treatment duration in regard to outcome is always raised. Mirski and colleagues reported a good recovery following 53 days of HDST, and suggested the following: no time limit when the potential prognosis is good, defined as a young individual with a healthy premorbid state, a self-limited and possibly reversible disease process, and when neuroimaging shows no radiographic lesion that suggests a poor prognosis, such as laminar necrosis. Again, the etiology is an important determinant of outcome, and therefore, an extensive evaluation for an underlying cause must be done, especially for infectious or metabolic disorders.

CLINICAL PEARLS

1. Refractory SE is defined as SE that persists despite appropriate initial antiepileptic drug treatment.
2. It is mandatory to look for an underlying cause in RSE. The term "metabolic" usually refers to electrolyte disorders, whereas the term "inborn error of metabolism" refers to a metabolic or genetic disease.
3. Aiming therapy to a burst-suppression EEG is controversial. It is not clear if clinical or electrographic seizure suppression has the same efficacy as achieving burst-suppression.
4. If seizures recur after discontinuing HDST, treat with repeat HDST cycles, but only if a good outcome is possible.
5. It is highly unlikely that an acute symptomatic, progressive encephalopathy, or a remote symptomatic case with an acute precipitant, will ever return to baseline functioning, and therefore, ethical considerations must be employed when making decisions about prolonged HDST.

SUGGESTED READING

Claassen J, Hirsch LJ, Emerson RG, Mayer SA. (2002) Treatment of refractory status epilepticus with pentobarbital, propofol, or midazolam: a systematic review. *Epilepsia* 43: 146–53.

Mirski MA, Williams MA, Hanley DF. (1995) Prolonged pentobarbital and phenobarbital coma for refractory generalized status epileptics. *Crit. Care Med.* 23: 400–404.

Sahin M, Menache C, Holmes GL, Riviello JJ. (2001) Outcome of severe refractory status epilepticus in children. *Epilepsia* 42: 1461–1467.

Sahin M, Menache C, Holmes GL, Riviello JJ. (2003) Prolonged treatment for acute symptomatic refractory status epilepticus: outcome in children. *Neurology* 61: 398–401.

Shorvon S. (1994) *Status Epilepticus: Its Clinical Features and Treatment in Children and Adults.* Cambridge, U.K. Cambridge University Press.

Van Gestel JP, van Oud-Alblas HJ, Malingre M, Ververs FF, Braum KP, van Nieuwenhuizen O. (2005) Propofol and thiopental for refractory status epilepticus in children. *Neurology* 65: 591–592.

19 Myoclonus Epilepsy with Ragged Red Fibers

Russell P. Saneto, D.O., Ph.D.

CONTENTS

CASE PRESENTATION

Our patient is a previously healthy 23-year-old, right-handed woman who started having frequent, repetitive jerking at 13 years of age. These movements were described as lightening-like myoclonia, mostly occurring during the day but not particularly in the early morning hours. As time progressed, she also developed myoclonic seizures, as well as infrequent generalized tonic–clonic seizures. Her parents described that she had become more "clumsy" over the past several years. Clumsiness was attributed to both muscle weakness as well as myoclonia. Initially, it was thought that drop seizures were occurring, but on further testing, most falling episodes were due to ataxia and myoclonia. On neurological exam, she displayed cerebellar ataxia and hypotonia. Once an excellent student, she began to display a progressive regression in cognitive functioning as myoclonia intensified. Over time, she developed optic atrophy; however, other abnormalities were not found.

The electroencephalogram (EEG) predominantly showed bursts of atypical generalized spike-and-wave with a disorganized slow background (Figure 19.1). There were also independent spike discharges over the left and right hemisphere, and diffuse slow delta bursts. Over several years, the background slowing has been persistent but not progressive. There were also photomyoclonic responses with photic stimulation. On repeat EEG, massive body myoclonia was demonstrated (Figure 19.2). Initial magnetic resonance imaging

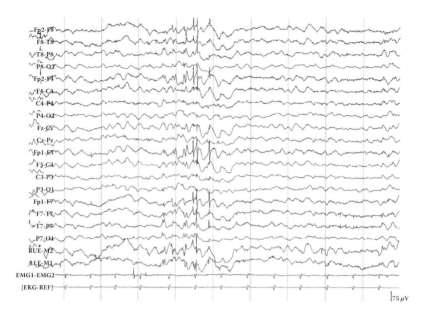

FIGURE 19.1   This EEG epoch shows atypical generalized spike-and-wave discharges. There is a disorganized slow background.

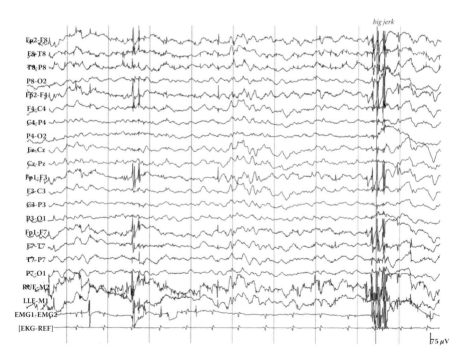

FIGURE 19.2   This EEG epoch demonstrated two events of massive body myoclonia without significant EEG change. There is persistence of the disorganized slow background with frontal predominant narrow spikes.

(MRI) of the brain at the time of diagnosis was interpreted as normal. A repeat scan a year later was also read as normal.

Muscle biopsy showed ragged red fibers on histological staining. Electron transport chain enzymology demonstrated normal activity. However, molecular testing identified an A8344G mutation in the mitochondrial DNA.

DIFFERENTIAL DIAGNOSIS

The differential diagnosis in a child/teenager who develops progressive neurological regression, ataxia, and myoclonic epilepsy includes the progressive myoclonic epilepsies. Early on in the evaluation process, the hereditary ataxia syndromes would be a consideration if the ataxia component was predominant. However, the presence of myoclonia and myoclonic seizures would be more consistent with the progressive myoclonic epilepsies. There are four main elements of the progressive myoclonic epilepsies: (1) myoclonic jerks that are segmental, fragmentary, and erratic in region; (2) epileptic seizures—mainly generalized tonic–clonic and massive myoclonic seizures; (3) progressive mental deterioration; and (4) variable neurological signs and symptoms—mainly cerebellar, extrapyramidal, and action myoclonus. Most are genetically determined, and all have a neurologically degenerative course. The signs and symptoms are usually specific or highly suggestive of a particular type of epilepsy. The typical progressive myoclonic epilepsies include: Unverricht–Lundborg disease, myoclonus epilepsy with ragged red fibers (MERRF), Lafora body disease, Sialidosis type 1, neuronal ceroid lipofuscinoses, juvenile neuronopathic Gaucher disease, dentatorubral-pallidoluysian atrophy, and juvenile neuroaxonal dystrophy. Most feel that the concept of a definite syndrome of progressive myoclonic epilepsy is archaic, as many epileptic syndromes may have transient episodes of ataxia and/or mental regression and are not genetic. However, the term *progressive myoclonic epilepsy* has been maintained in the new guidelines for classification of seizures and syndromes.

We used the precise term *myoclonus epilepsy with ragged red fibers*. Although *myoclonic* seizures are seen, there are also multiple myoclonus events, and therefore, the name of the mitochondrial disease reflects these three characteristics. The term *myoclonic epilepsy with ragged red fibers* is often used as well. We prefer to use the former due to the more complete definition of the disorder.

DIAGNOSTIC APPROACH

The age of presentation, associated clinical signs and symptoms, clinical course, pattern of inheritance, and ethnic origin of the patient are invaluable in diagnosis of this group of epilepsies. MERRF has a variable age of onset from 3 years to adulthood. This wide range of onset can be confusing due to other myoclonic epilepsies beginning at various ages, many of which are benign. Usually, the clinical history, EEG, and serial neurological examination will help differentiate the various myoclonic epilepsies. There are also other types of mitochondrial disease that express myoclonic seizures, developmental stagnation or regression, and various organ involvements.

Based on the clinical presentation and progression of cognitive loss, associated symptoms, and ethnic origin, most of the other progressive myoclonic epilepsies can be potentially diagnosed. Other than the canonical features of myoclonus, generalized seizures, ataxia, and ragged red fibers in muscle, there are frequent other clinical abnormalities noted in MERRF, including sensorineural hearing loss, peripheral neuropathy, short stature, exercise intolerance, and optic atrophy. Less frequent clinical signs reported are cardiomyopathy, preexcitation arrhythmia (Wolf–Parkinson–White), pigmentary retinopathy, ophthalmoparesis, pyramidal signs, and multiple lipomas.

The majority of mitochondrial diseases are multisystem disorders, with those organs requiring the most energy usually demonstrating the presenting phenotype. Myoclonus, generalized seizures, and normal early development are typical in the diagnosis of MERRF. Other mitochondrial diseases due to respiratory chain defects and different mitochondrial DNA mutations can present similarly, confusing the correct diagnosis. Screening labs of serum lactate, quantitative serum amino acids, serum acyl carnitine profile, and quantitative urine organic acids should begin to differentiate possible MERRF from other progressive myoclonic epilepsies, as well as other mitochondrial diseases. Unlike most respiratory chain disorders, the EEG shows a generalized spike/polyspike pattern in MERRF. Nuclear magnetic resonance (MRI) and proton magnetic resonance spectroscopy (MRS) imaging may also help differentiate possible diagnoses. Voxels over the CSF space and brain showing a lactate peak on MRS would suggest the possibility of a mitochondrial disease. MRI findings can be useful in differentiation of Leigh syndrome and mitochondrial encephalomyopathy, lactic acidosis, and stroke-like episodes (MELAS) with MERRF. MRI in MELAS often demonstrates areas of abnormal T2 signal suggestive of ischemia, whereas Leigh syndrome is associated with abnormal T2 signal in the brainstem and basal ganglia suggestive of necrosis. If a strong indication of maternal inheritance is present, clinical history is compatible, and there are lactic acid elevations in CSF and serum, as well as bland findings in other biochemical testing, the investigation of a possible gene mutation in the mitochondrial DNA could be pursued at this point.

If the accrued evidence is indicative but not defining, then proceeding to muscle biopsy for further analysis would be suggested. The finding of ragged red fibers using Gomori trichrome staining would demonstrate the fourth feature of MERRF. Ragged red fibers in the child and adolescent are very unusual and, if present, would be confirmatory, given the presence of the other three features. Often, there are cytochrome oxidase negative fibers in both ragged red fibers as well as nonragged red muscle fibers. If respiratory chain enzymology is performed, deficient enzyme activity may or may not be found. It is important that, if respiratory chain abnormalities are found and the clinical suspicion is MERRF, ongoing testing should continue. Genetic testing to evaluate for commonly associated mutations should be undertaken to confirm the diagnosis for genetic counseling for other siblings and family members. We have found patients with many phenotypic qualities of MERRF, but without mitochondrial DNA mutations or ragged red fibers, who demonstrate electron transport chain deficiencies.

Molecular gene testing would be the next step. If muscle tissue is available, it would be the preferred tissue for examination, but isolated leukocytes can also be used for testing. The most common mitochondrial DNA mutation associated with MERRF is in the gene *MT-TK* encoding tRNA[Lys]: A8344G. Although over 80% of affected patients have the A8344G mutation, another 10% of patients have other point mutations within the *MT-TK* gene: T8356C, G8363A, and G8361A. There are also other rare mutations in the *MT-TK* gene as well as other mitochondrial coded proteins. Mutation in the complex I subunit MT-ND5 has been found to cause an overlap syndrome with MERRF and MELAS phenotype.

TREATMENT STRATEGY

Treatment with conventional seizure medications may reduce seizures initially, but will seldom produce complete remission as the disease progresses. Those medications that are usually used for myoclonic seizures generally are more efficacious: valproic acid, lamotrigine, zonisamide, levetiracetam, and benzodiazepines. However, medication efficacy tends to be patient specific, and no prospective studies have been performed. In a small study of five patients with mitochondrial disease due to respiratory chain dysfunction, vagus nerve stimulation did not produce reduction in myoclonic seizure frequency. This suggests that placement of the vagus nerve stimulator device be undertaken with caution.

The addition of L-carnitine and coenzyme Q10 has been advocated by some to improve mitochondrial function. However, no prospective studies have been performed to support this assertion. Standard pharmacologic therapy is used to treat other specific organ involvement, such as cardiac symptoms. Currently, there is no treatment for the genetic defect.

LONG-TERM OUTCOME

The outcome in patients with MERRF depends somewhat on heteroplasmy. Heteroplasmy is based on the idea that there are many mitochondria per cell, some of which may contain the mutation whereas others do not. Those patients having a higher percentage of the mutation likely express the disease earlier in life and have a more progressive course. Heteroplasmy may also account for the variation in disease expression in maternal relatives. Tissue distribution also plays a part in outcome; as more organ systems become involved, there is an increased compromise to the quality of life as well as longevity.

PATHOPHYSIOLOGY/NEUROBIOLOGY OF DISEASE

The molecular pathogenesis of mitochondrial tRNA[Lys] mutations is not completely understood. However, experiments using rho° cell lines have unveiled important clues. Rho° cell lines are permanent human cell lines emptied of their mtDNA by exposure to ethydium bromide, then repopulated with mitochondria harboring specific mutations. These transmitochondrial cybrids with a high mutational load have correlated with decreased protein synthesis, and reduced oxygen consumption and respiratory chain function. The 8344 mutation has also been shown to cause impairment of mitochondrial translation in cultured myoblasts. Polypeptides containing higher numbers of lysine residues are more severely affected by the tRNA^{Lys} mutation, thus suggesting a direct inhibition of protein synthesis. Furthermore, the 8344 mutation has been associated with defects in aminoacylation capacity as well as lower steady-state levels of tRNA^{Lys}. What remains unclear is how these defects orchestrate MERRF pathogenesis.

CLINICAL PEARLS

1. Evolving multisystemic organ system involvement in a previous normal patient with progressively medically resistant myoclonus should raise suspicion about MERRF.
2. There is no clear correlation between genotype and clinical phenotype for affected individuals, so clinical judgment is of utmost importance.
3. If clinical suspicion is MERRF but leukocyte testing is negative, other tissues (in particular muscle) should be used for detection of mutations.
4. Medications effective against myoclonus, such as benzodiazepines and levetiracetam, form the cornerstone of treatment but may fail to control seizures as the disease progresses.

SUGGESTED READING

Berkovic SF, Cochius J, Andermann E, Andermann F. (1993). Progressive myoclonus epilepsies: clinical and genetic aspects. *Epilepsia* 34, S19–S30.

DiMauro S, Davidzon G. (2005). Mitochondrial DNA and disease. *Ann. Med.* 37, 222–232.

Fukuhara N, Tokiguchi S, Shirakawa K, Tsubaki T. (1980). Myoclonus epilepsy associated with ragged red fibers (mitochondrial abnormalities): disease entity or a syndrome? Light- and electron-microscopic studies of two cases and review of literature. *J. Neurol. Sci.* 47: 117–133.

Hammans SR, Sweeney MG, Brockington M. et al. (1993). The mitochondrial DNA transfer RNA (Lys)A > G (8344) mutation and the syndrome of myoclonic epilepsy with ragged red fibers (MERRF): relationship of clinical phenotype to proportion of mutant mitochondrial DNA. *Brain* 116 (Pt. 3), 617–632.

Shoffner JM, Lott MT, Lezza AM, Seibel P, Ballinger SW, Wallace DC. (1990). Myoclonic epilepsy and ragged red fiber disease (MERRF) is associated with a mitochondrial DNA tRNA(Lys) mutation. *Cell* 61: 931–937.

20 Sturge–Weber Syndrome

Asit K. Tripathy, M.D. and Ajay Gupta, M.D.

CONTENTS

CASE PRESENTATION

A 9-month-old male, product of a nonconsanguineous marriage, was seen for management of intractable epilepsy. The pregnancy was unremarkable except for maternal supraventricular tachycardia (previous history of similar episodes). A left facial nevus was noted at birth. At the age of 6 weeks, parents noticed episodes of whole-body stiffness, arching, and upward eye rolling, followed by vomiting. He would become limp and lethargic for several minutes after the spells. Gastroesophageal reflux was suspected but medical management proved unsuccessful. At the age of four months, a nocturnal episode of irritability, pallor, and vomiting lasting several hours was followed by right-sided hemiplegia, for which he was hospitalized. Ischemic stroke was suspected; however, an acute noncontrast brain computed tomography (CT) was normal. The right-sided hemiplegia gradually recovered over 4–6 weeks without residual weakness. Subsequently, parents noted new episodes of behavioral arrest, body stiffness, a dusky color, unresponsiveness, and right-foot jerking for 1–2 minutes. He would become limp, lethargic, and have right-arm weakness for several minutes after each spell. The spells occurred approximately once a day. Once every two weeks, this type of seizure would evolve into a generalized motor seizure. His seizures were treated with phenobarbital, phenytoin, oxcarbamazepine, and clonazepam without any success. At 8 months of age, his parents noticed left-hand preference, and concerns for developmental delay were raised. Physical examination was remarkable for port-wine nevus in the trigeminal V1 distribution involving the left upper eyelid and medial canthus. Dexterity was impaired,

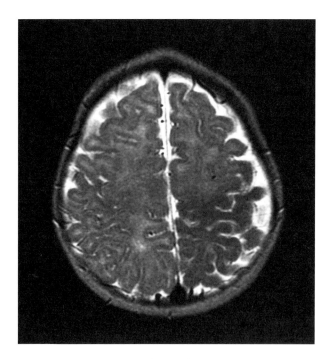

FIGURE 20.1 T2-weighted brain MRI of the patient showing volume loss in the left parietal and occipital region involving both gray and white matter.

and weakness was noted in the right hand and arm, especially when he tried to approach or transfer objects from the left hand, suggesting moderate right hemiparesis. Muscle tone, bulk, strength, and reflexes were symmetrical on both sides. Video-electroencephalogram (EEG) monitoring revealed interictal sharp waves in the left parieto-occipital region, with continuous slowing and decreased background rhythm in the left hemisphere. Ictal EEG showed onset from the left parieto-occipital region during a typical complex partial seizure, ending in a right hemiclonic seizure. Brain magnetic resonance imaging (MRI) showed a typical finding of Sturge–Weber syndrome (SWS; Figure 20.1). Brain flurodeoxyglucose–positron emission tomography (FDG-PET) showed hypermetabolism in the left posterior quadrant, suggesting increased FDG uptake due to a nearly continuous burst of interictal spiking. Ophthalmic examination revealed a likely right visual field deficit by confrontation testing; intraocular pressure and fundus examination were normal. After discussion of risks, benefits, and alternatives, the patient underwent functional hemispherectomy at age 9 months. There was no further recurrence of seizures, and antiepileptic medications were discontinued 10 months after the surgery. At a 5-year follow-up, the patient remains seizure free and has mild developmental delay. He is ambulatory, with his left hand being weak and spastic.

DIFFERENTIAL DIAGNOSIS

The most important entity to consider in the differential is facial venous angioma without any cerebral angiomatosis. There are other rare congenital vascular disorders involving brain and skin. Klippel–Trénaunay syndrome classically presents as a triad of varicosities, bone or soft-tissue hypertrophy, and cutaneous hemangiomas. Wyburn–Mason syndrome is a congenital neurocutaneous entity comprised of ipsilateral arteriovenous malformations of the midbrain, vascular abnormalities affecting the visual pathway, and facial nevi. PHACE syndrome comprises of posterior fossa brain malformations, hemangiomas, arterial anomalies, coarctation of the aorta, cardiac defects, and eye abnormalities.

DIAGNOSTIC APPROACH

The clinical features are variable, but the association of neurological deficits and port-wine stain of the face suggest SWS. SWS occurs sporadically in all races. The prevalence is estimated to be 1 per 50,000. The dermatological lesion of a facial port-wine stain (PWS) is usually present at birth, and consists of a flat lesion of variable size involving the upper eyelid and forehead. The size of the cutaneous angioma does not predict the size of the intracranial angioma. It is unilateral in 70% of the cases, usually ipsilateral to the brain involvement. Even when the facial angioma is bilateral, the pial angioma tends to be unilateral or highly asymmetric in most patients. Children with a PWS involving the first branch of the trigeminal nerve have a 25% increased risk of SWS. The characteristic neurological and radiographic features of SWS may rarely be present without cutaneous angioma. Only 10–20% of children with a port-wine nevus of the forehead have a leptomeningeal angioma. Typically, SWS involves the occipital and posterior parietal lobes, but it can affect other cranial regions and both cerebral hemispheres. Bilateral brain lesions occur in 15% of children.

Seizures are the most common neurological presentation and occur in 72–80% of children with SWS. The age range of seizure onset varies between birth and 23 years, with median age being 6 months. The risk of developing seizures is highest in the first two years of life and occurs earlier in patients with bilateral disease. The most common type of seizure is a partial seizure, usually with a hemitonic or hemiclonic semiology. Secondarily generalized seizures are commonly seen, usually later in childhood and in adolescents. There is also an increased incidence of prolonged seizures or status epilepticus in SWS patients. Fever and infection may trigger the onset of seizures in many children.

Seizures frequently accompany stroke-like episodes. Onset of a motor deficit may precede a cluster of prolonged seizures rather than seizures followed by Todd's paralysis; however, this distinction is difficult to make in children. Fixed hemiparesis contralateral to the facial angioma eventually occurs in 50% of children. It often appears after a focal-onset seizure and progresses in severity in a stuttering fashion after subsequent seizures. Transient episodes of hemiplegia, not related to clinical or EEG evidence of seizure activity, may also occur. Some patients have associated migraine-like headache, attention deficit disorder, and mental retardation. Glaucoma occurs in

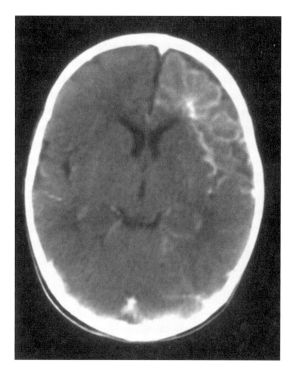

FIGURE 20.2   Noncontrast CT head of a patient with SWS with typical gyriform calcification in the left frontal and parietal regions. Calcification may not be appreciated on routine brain MRI and can appear later in life.

30–70% cases, and usually develops before the age of 10 years. Presence of vascular malformation in the distribution of the V1 segment increases the probability of glaucoma. Bupthalmus and ambylopia are present in some newborns. There may be an associated vascular abnormality in the conjuctiva, sclera, retina, and choroid. There is also increased incidence of retinal detachment secondary to hemorrhage from choroidal vessels. Eye involvement may result in acute or chronic visual loss that may not be readily apparent in a young child without an evaluation by an ophthalmologist.

Children born with PWS covering the trigeminal V1 distribution should have a contrast-enhanced brain MRI. The imaging shows enhancement of the leptomeningeal angioma, enlarged transmedullary veins, choroid plexus hypertrophy, white matter abnormalities, patchy parenchymal gliosis, calcification, neuronal loss, and gliosis. However, the brain MRI may only show subtle or no abnormalities in young infants who are subsequently diagnosed with SWS. CT scanning of the brain, although not routinely done, may show cortical calcifications, typically described as "tram track" or "gyriform" appearance (Figure 20.2). Calcification may be absent or minimal in neonates and infants. Functional imaging with FDG-PET often demonstrates cortical hypometabolism. Another emerging MR technique, diffusion tensor imaging (DTI), often demonstrates abnormal water diffusion, suggesting a lack of integrity of the white matter underneath the leptomeningeal angioma. The EEG frequently shows ipsilateral slowing to the cerebral involvement with or without spike and sharp waves. Quantitative EEG (qEEG) may provide an objective measure of EEG asymmetry that correlates with clinical status and brain asymmetry seen on MRI.

TREATMENT STRATEGY

Seizures may be difficult to control with antiepileptic medications. Broad-spectrum antiepileptic medications effective against partial seizures may prove helpful. In some patients, the disease appears to be progressive, and there is a view that early resective surgery may be effective in halting the progression. It is not possible to predict who will develop medically intractable epilepsy. Surgery should be considered when seizures are refractory to medical treatment. Visually guided complete excision of the angiomatous cortex with or without the guidance of electrocorticography is the primary surgical procedure for epilepsy surgery. Hemispherectomy is generally considered in children with extensive unilateral brain involvement and a fixed hemiparesis. The ketogenic diet or vagus nerve stimulation may provide alternative treatment options for refractory patients, particularly in those with bilateral ictal onset zones. Aspirin 3–5 mg/kg/day is often recommended, with SWS, as primary prevention or secondary prevention after the first stroke-like episode, but the literature is mixed about its utility, and there have been no controlled trials. Laser therapy is the recommended intervention for cutaneous PWS. Multiple treatments are usually required to significantly lighten PWS lesions. PWS may grow and thicken as the child grows. Medical and surgical treatment of glaucoma includes beta blockers, carbonic anhydrase ophthalmic drops, and surgery. Regular evaluation by an ophthalmologist is recommended, particularly for patients with choroidal lesions.

LONG-TERM OUTCOME

The clinical progression of SWS is characterized by a stuttering course with periodic worsening, and episodes of status epilepticus and stroke-like episodes. There is a great risk for neurologic complications in widespread or bihemispheric disease. Seizures occurring before 2 years of age increase the risk of mental retardation and refractory epilepsy. Some patients continue to have daily seizures after the initial deterioration, despite various antiepileptic medications, whereas others have long seizure-free intervals. The timing of surgery is important. The majority (70–80%) of patients may be seizure free or significantly improved (i.e., >75–90% seizure reduction) after surgery, and early surgery may improve developmental outcome in refractory patients. The completeness of resection or disconnection of diseased tissue is an important factor in achieving epilepsy control.

PATHOPHYSIOLOGY/NEUROBIOLOGY OF DISEASE

SWS is a sporadic disease presumed to be caused by somatic mutation. During the sixth week of intrauterine life, the primitive embryonal vascular plexus develops around the cephalic portion of the neural tube and under the ectoderm, in the region destined to be the facial skin. In SWS, it is hypothesized that the vascular plexus fails to regress, as it should in the embryo in the 9th week, resulting in angiomatosis of related tissues. The intracranial lesion is thought to be due to proliferation of leptomeningeal vessels in the subarachnoid space, which causes shunting of blood away from the brain tissue. This shunting results in decreased blood flow, decreased venous return, and focal hypoxia leading to cellular death. This is seen radiographically as gliosis, volume loss, and calcification.

CLINICAL PEARLS

1. SWS patients with early onset and frequent seizures usually have a more severe clinical course. SWS may show progressive clinical deterioration with intractable epilepsy, neurological deficits, and cognitive regression.
2. Epilepsy surgery is an effective treatment for patients with refractory epilepsy. Epilepsy surgery should be considered early in the course of the disease.
3. When a child is born with a facial PWS involving upper and lower eyelids, contrast-enhanced brain MRI should be considered.
4. Eye involvement may result in acute and chronic visual loss; therefore, ophthalmic examination and follow-up examination by an expert is crucial to prevent loss of vision.
5. Aspirin is used as a primary and secondary prevention measure for stroke-like episodes, though its efficacy remains uncertain.

SUGGESTED READING

Bourgeois M, Crimmins DW, de Oliveira RS. et al. (2007). Surgical treatment of epilepsy in Sturge–Weber syndrome in children. *J. Neuosurg.* 106(1 Suppl.): 20–8.

Chugani HT, Mazziotta JC, Phelps ME. (1989). Sturge-Weber syndrome: a study of cerebral glucose utilization with positron emission tomography. *J. Pediatr.* 114(2): 244–53.

Di Rocco C, Tamburrini G. (2006). Sturge–Weber syndrome. *Childs. Nerv. Syst.* 22(8): 909–21.

Hatfield LA, Crone NE, Kossoff EH. et al. (2007). Quantitative EEG asymmetry correlates with clinical severity in unilateral Sturge–Weber syndrome. *Epilepsia* 48(1): 191–5.

Kossoff EH, Buck C, Freeman JM. (2002). Outcome of 32 hemispherectomies for Sturge–Weber syndrome worldwide. *Neurology* 59(11): 1735–8.

Sujansky E, Conradi S. (1995). Sturge–Weber syndrome: age of onset of seizures and glaucoma and the prognosis for affected children. *J. Child. Neurol.* 10(1): 49–58.

Thomas-Sohl KA, Vaslow DF, Maria BL. (2004). Sturge–Weber syndrome: a review. *Pediatr. Neurol.* 30(5): 303–10.

Section 4

The Child

21 Benign Epilepsy of Childhood with Central-Temporal Spikes

Paul M. Levisohn, M.D.

CONTENTS

CASE PRESENTATION

A 6-year-old boy presented with a prolonged generalized convulsion occurring shortly after the child went to bed. The onset was not observed. He was transported to a local emergency department, where his seizure was controlled with intravenous lorazepam. A computed tomography (CT) scan of the head in the emergency room was unremarkable. Discussion with the family revealed a past history of several nocturnal events with right facial and tongue clonic movements with drooling, lasting less than 1 minute. Electroencephalography (EEG) performed the following day showed prominent epileptiform activity with high-amplitude sharp waves with following slow waves, predominantly in the left central-temporal regions but occasionally seen independently on the right. The spike activity was activated (increased in frequency) by light sleep. A diagnosis of benign epilepsy of childhood with central-temporal spikes (BECTS), also known as rolandic epilepsy, was made. After discussion with the family, treatment with oxcarbazepine was initiated. The patient was seizure free on oxcarbazepine monotherapy for 1½ years, at which point the medication was tapered off. The patient has remained seizure free.

DIFFERENTIAL DIAGNOSIS

Clinical semiology with simple partial seizures suggests the presence of a focal structural abnormality. In particular, simple partial seizures may be due to malformations
of cortical development and congenital tumors such as dysembryoplastic neuroepithelial tumor (DNET). The occurrence of arousal with abnormal movements may
suggest parasomnias such as somnambulism. Unilateral central-temporal spikes
(CTS) on the EEG may support the diagnosis of cryptogenic localization-related
epilepsy that is either temporal or extratemporal. Nocturnal convulsive seizures may
suggest frontal-lobe epilepsy and may be confused with parasomnias. (See Chapter
37 on ADNFLE.) Likewise, if the only observed manifestation of the seizures is
generalized convulsive activity, the diagnosis of primary generalized epilepsy may
be erroneously entertained.

DIAGNOSTIC APPROACH

BECTS accounts for 13–23% of all epilepsy in children. The diagnosis of BECTS
will be strongly suggested by the clinical semiology of the seizures, with sensorimotor seizures without cognitive impairment (simple partial seizures). Typically, the
seizures affect facial musculature and may be associated with aphasia or dysarthria,
often with drooling. The seizures are most likely to occur shortly after sleep onset
or before awakening. Diurnal seizures occur in approximately 46% of children. Secondary generalization is not uncommon, and is seen in nearly 50% of patients. Status
epilepticus has also been reported, though it is very unusual. BECTS is a childhood
disorder with the most common age of onset at 7–8 years and within a range of
6 months to 14 years. Postpubertal onset essentially excludes the diagnosis. Many
children will experience only a single seizure with three-quarters experiencing five
or fewer.

A normal developmental history and normal examination support the diagnosis,
but abnormal findings should not exclude the diagnosis. In addition to seizure semiology, the diagnosis of BECTS rests on the presence of typical EEG findings. The
interictal EEG will have a normal background. Epileptiform activity consisting of
high-voltage diphasic or surface-negative spikes with following slow wave is usually
present and is usually activated by light sleep. As the syndrome's name indicates,
the epileptiform activity is typically midtemporal and/or central. A horizontal dipole
may be noted that is negative in the central-temporal region and positive frontally.
(See Figures 21.1 and 21.2.) Atypical spike localization may be seen. Bilaterally independent epileptiform activity is seen in somewhat less than half of the patients. Typically, there is marked activation of epileptiform activity in light sleep, and therefore
a sleep-deprived EEG (preferably with natural sleep) should be obtained, especially
if the waking state EEG is normal. Generalized spike–wave discharges may be seen
in drowsiness and light sleep, and should not discourage the diagnosis.

The presence of CTS in the absence of seizures does not establish the diagnosis of BECTS. CTS are seen in up to 3.5% of normal children without a history of
seizures and do not in themselves indicate a seizure diathesis. CTS may be seen in
children with other neurologic disorders such as ADHD as well as in other forms of

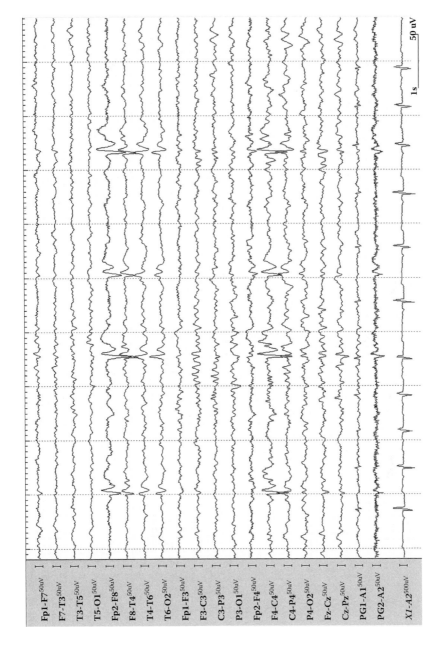

FIGURE 21.1 This EEG demonstrates right centrotemporal spikes on a bipolar montage.

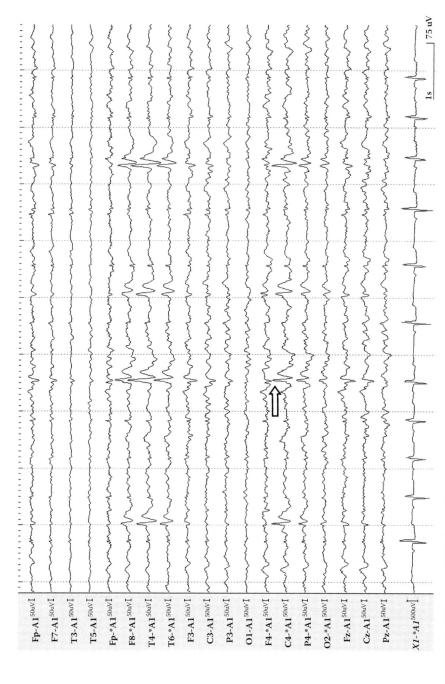

FIGURE 21.2 The same EEG data as in Figure 21.1 shown on a reference montage to A1. Note the small positivity at Fp2 and F4 consistent with a horizontal dipole (open arrow).

epilepsy. They occur in a significant proportion of neurologically normal siblings of children with BECTS, supporting an autosomal dominant mode of inheritance for the EEG abnormality but not for the epileptic disorder itself.

For the patient with classic features of BECTS, including normal examination, typical seizures, and CTS, neuroimaging is not usually indicated. Abnormal medical and developmental history or examination in children with typical features of BECTS do not exclude the diagnosis and do not predict a more ominous course. Some children will demonstrate abnormalities on neuropsychological testing, especially relating to language.

Children with atypical clinical or EEG features should have imaging performed because treatable CNS abnormalities may be present, including malformations of cortical development. Rarely, children with what appears to be typical BECTS at onset may experience a more disabling course, developing other seizure types, becoming medication resistant, and experiencing progressive neurocognitive dysfunction.

Video-EEG monitoring is generally not of value in establishing the diagnosis. Typically, seizures do not occur with sufficient frequency to be recorded with any confidence. Light sleep can almost always be recorded in an appropriate outpatient setting. Nocturnal polysomnography may be suggested by the occurrence of events during sleep, but careful history and typical interictal EEG will rule out a primary sleep disorder. Although magnetoencephalography (MEG) has been employed for investigation of the pathophysiology of BECTS, MEG has no demonstrated clinical value in this setting.

Reports regarding neuropsychological abnormalities in children with BECTS have suggested that neuropsychological testing may be of value in such patients. However, in the absence of demonstrated learning and cognitive difficulties, such assessments need not be performed routinely.

TREATMENT STRATEGY

The natural history of BECTS supports the common practice of withholding medication treatment in the child with the disorder who has experienced only a single or few nocturnal seizures. However, frequent seizures, diurnal seizures, and secondary generalization may be indications for initiating treatment to prevent recurrence. There is a single published randomized controlled trial regarding antiepileptic drug (AED) treatment of BECTS, demonstrating that sulthiame is effective. However, sulthiame is not available in the United States. A randomized controlled trial of gabapentin has been reported in abstract only and supports its efficacy in the disorder. Clinical experience is greatest with carbamazepine, but other medications suitable for treatment of partial seizures are also used. Low doses of medications appear to be effective, and some practitioners use once-per-day dosing prior to bedtime to reduce side effects. No studies have specifically addressed the question of duration of treatment. Common practice with other idiopathic epilepsy syndromes suggest that treatment for more than 2 years seizure free is not necessary, and that a shorter duration may be sufficient.

LONG-TERM OUTCOME

The prognosis for BECTS is excellent, even in children with frequent seizures. Essentially, all children become seizure free by mid-adolescence, although the EEG abnormalities may persist somewhat longer. Thus, >98% of children with the diagnosis of BECTS will remit and will remain free of seizures on long-term follow-up. As noted earlier, however, some children develop an atypical syndrome and may experience poorly controlled seizures and neuropsychological dysfunction.

PATHOPHYSIOLOGY/NEUROBIOLOGY

BECTS is classified as an idiopathic epilepsy (implying genetic), but well-designed studies have failed to define the genetics of the disorder. The mode of inheritance is unclear and is likely multifactoral, although reduced penetrance of an autosomal dominant disorder is possible. The EEG pattern of CTS appears to be inherited as an autosomal dominant trait. As noted, CTS occur with significant frequency in children without a seizure disorder. Source localization of EEG epileptiform activity suggests that epileptiform activity is multifocal in origin.

CLINICAL PEARLS

1. The clinical prognosis of BECTS is excellent, although variations of the disorder are seen. It may be difficult to provide a confident good prognosis at the onset of the disorder if atypical clinical presentation is seen. Thus, abnormal development, the occurrence of other seizure types, and atypical EEG patterns suggest the possibility of a less optimistic outcome.

2. Recording stage II sleep is necessary to demonstrate the presence or absence of the typical pattern of rolandic spikes. A sleep-deprived EEG is critical in the diagnosis of BECTS.

3. Although most children with BECTS are otherwise normal, there is a risk of learning difficulties, especially language-related, in children with the disorder. Formal assessment of learning may be appropriate in some children.

4. Although optimal duration of treatment, when initiated, is unclear, most clinicians will treat for two years seizure free. The persistence of CTS is not uncommon but need not alter the decision to discontinue treatment.

5. Despite the lack of studies providing adequate evidence for determining which medications to use in those who are felt to be candidates for antiepileptic medication, carbamazepine and oxcarbazepine are often used when treatment is indicated.

SUGGESTED READING

Bali B, Kull LL, Strug LJ. et al. (2007) Autosomal dominant inheritance of centrotemporal sharp waves in rolandic epilepsy families. *Epilepsia* 48, 2266–2272.

Bouma PAD, Bovenkerk AC, Westendorp RGJ, Brouwer OF. (1997) The course of benign partial epilepsy of childhood with centrotemporal spikes: a meta-analysis. *Neurology* 48, 430–437.

Kellaway P. (2000) The electroencephalographic features of benign centrotemporal (rolandic) epilepsy of childhood. *Epilepsia* 41, 1053–1056.

Metz-Lutz M, Filippini M. (2006) Neuropsychological findings in rolandic epilepsy and Landau–Kleffner syndrome. *Epilepsia* 47(Suppl. 2), S71–S75.

Vadlamudi L, Harvey AS, Connellan MM. et al. (2004) Is benign rolandic epilepsy genetically determined? *Ann. Neurol.* 56, 129–32.

22 Childhood Absence Epilepsy

L. Matthew Frank, M.D.

CONTENTS

CASE PRESENTATION

Harriet's family had become concerned about her. At 7 years she had been an honor role second grader, yet recently her grades were slipping. Her family had also begun to receive notes sent home from her teachers, as she was no longer attentive during class. Harriet grew increasingly cranky and unhappy. She was at a loss to explain her deteriorating school performance. It was not until one evening at the dinner table, when Harriet was telling a story, that her parents noticed that, when she paused, she had slight ptosis with slight rhythmic blinking and mouthing movements that did not abate when they called her by her name. This unresponsiveness lasted about 20 seconds, after which Harriet had no recollection of her parent's alerting efforts.

Chagrinned that in blaming school performance on psychological issues they may have missed an important medical problem, her parents called their pediatrician who arranged for an electroencephalogram (EEG) and neurology consultation. The EEG demonstrated frequent 3 Hz spike-and-slow-wave discharges in awake and sleep states that were induced by hyperventilation (HV). Harriet was unable to repeat words said to her during her longer 3 Hz discharges. Motor automatisms were evident during these lengthier discharges. The frequency of spontaneous discharges on EEG suggested that Harriet was probably having a large number of unrecognized absence events daily because only the longer ones had obvious motor accompaniment. Her neurologist successfully started an antiepileptic drug (AED) and Harriet's attention and grades improved.

DIFFERENTIAL DIAGNOSIS

Absence seizures in childhood are often divided into two major categories: typical and atypical. The absences of childhood absence epilepsy (CAE) are considered "typical" and do not have accompanying myoclonus, generalized tonic–clonic or partial seizures, or arrhythmic epileptiform EEGs, any of which would make them "atypical." The staring spells occurring in neurologically impaired children are usually atypical.

There have been attempts to characterize the typical absences of CAE as simple or complex, depending on whether there are associated and more complex motor phenomena. The distinction between simple and complex has limited clinical value, because pharmacologic approaches and outcomes are virtually identical.

Once suspected, CAE is among the easiest, most straightforward, and satisfying diagnoses in neurology. Unfortunately, many children with CAE have a delayed diagnosis due to the subtlety and nonalarming nature of their presentation. The absences of CAE may occur relatively infrequently; they may be too short to be recognized; they may not have prominent automatisms or autonomic changes to aid in their detection; or the child may simply have relatively disinterested caretakers.

A diagnosis often confused with CAE is complex partial epilepsy having a short or absent postictal phase. The response to HV and the EEG will clearly distinguish these two seizure types. EEG is an essential part of the diagnostic effort, as some partial epilepsies and some clinical states (daydreaming) may be difficult to conclusively eliminate on clinical or historical grounds alone.

DIAGNOSTIC APPROACH

No other epilepsy allows such simple confirmation as an HV test done in the office. The induction of a state of unresponsiveness, halting of the HV effort, the appearance of minor automatisms, and the almost immediate return of consciousness and normal mentation without lethargy or the need for sleep is virtually diagnostic of uncontrolled CAE. An EEG will confirm that these clinical events are related to approximately 3 Hz rhythmic discharges and will also serve to confirm the lack of other EEG abnormalities (Figures 22.1 and 22.2). The EEG with HV is extremely sensitive for the diagnosis of CAE; a normal or negative EEG with HV in an untreated child with CAE is extremely rare. Photosensitivity is unusual in CAE. Adding to the clinical satisfaction of treating CAE is the observation that, in this form of epilepsy, there is a strong correlation between normalization of the EEG and AED efficacy. On the other hand, if there is any shortcoming in CAE management, it is that the decision as to when AEDs can be withdrawn cannot reliably depend on EEG findings unless the patient has been weaned from AEDs.

The clinician must recognize that it is not uncommon for a child with CAE to have fragmentary spike-and-slow-wave bursts during sleep or to have subtle asymmetries in the onset, distribution, or resolution of the 3 Hz discharges, which should not mitigate against the correct diagnosis. Similarly, a second or so of EEG slowing after a 3 Hz burst is quite consistent with CAE.

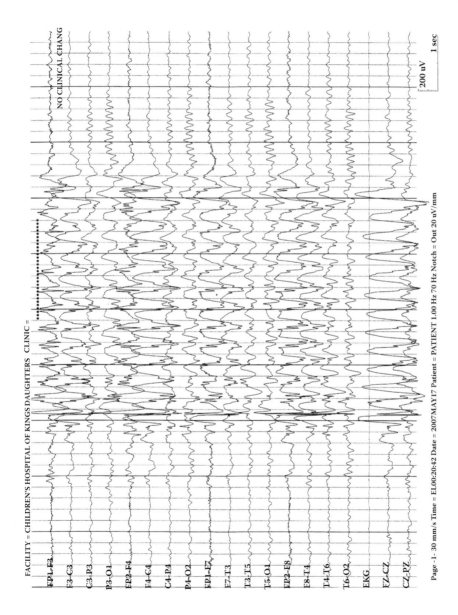

FIGURE 22.1 This EEG shows a spontaneous, approximately 3.5 Hz, frontally dominant generalized spike-and-slow-wave discharge of 4–5 seconds duration having a slightly asymmetric onset but becoming symmetric within 0.5 seconds. Note the well-developed 8 Hz posterior background before and after the discharge. There was no observed clinical change with this electrographic seizure.

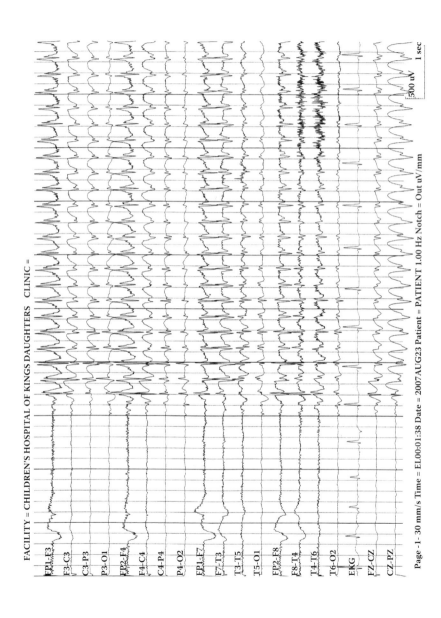

FACILITY = CHILDREN'S HOSPITAL OF KINGS DAUGHTERS   CLINIC =

FP1-F3
F3-C3
C3-P3
P3-O1
FP2-F4
F4-C4
C4-P4
P4-O2
FP1-F7
F7-T3
T3-T5
T5-O1
FP2-F8
F8-T4
T4-T6
T6-O2
EKG
FZ-CZ
CZ-PZ

Page -1- 30 mm/s Time = EL00:01:38 Date = 2007AUG23 Patient = PATIENT 1.00 Hz Notch = Out uV/mm

500 uV   1 sec

(A)

FIGURE 22.2A,B This EEG shows a spontaneous, continuous 13-second 3–4 Hz generalized discharge extending over two pages (10 sec/page), during which the patient had a clinical absence seizure. Background rhythms are not easily observable because the gains are reduced to allow visualization of the high-amplitude spike morphology. See calibration signal and compare to Figure 22.1.

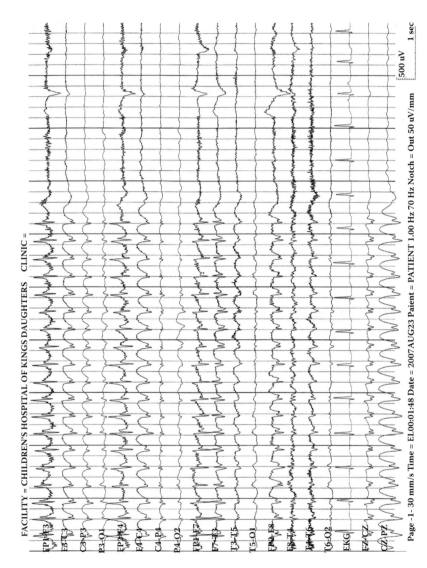

(B)

FIGURE 22.2A,B (continued)

Whether neuroimaging is necessary in typical CAE has been debated. If there is a suspicion on history that the seizures may not be typical for CAE, if the physical exam is not completely normal, if the EEG is not as expected, or if there is a worrisome family history, ordering magnetic resonance imaging (MRI) can not be faulted.

A discussion about CAE would not be complete without a few comments regarding JAE (juvenile absence epilepsy) because the age ranges of these two syndromes overlap and prognosis and treatment are different. JAE usually presents later than age 9, but there are rare CAE patients presenting that late and occasional JAE patients presenting earlier. JAE frequently evolves to include other generalized seizure types (generalized tonic–clonic and myoclonic) and often has an associated family history of epilepsy, suggesting a strong genetic basis. Many JAE patients evolve into JME (juvenile myoclonic epilepsy), and JAE treatment regimens must take into account this risk for other seizure types.

CAE usually resolves with the emergence of puberty, the same time that JAE is making its appearance. The child's Tanner stage may be helpful in deciding which diagnostic and therapeutic direction to take.

A child presenting to the emergency department with a prolonged confusional state for which no other explanation can be found may warrant an EEG to consider absence status epilepticus.

The automatisms occurring with CAE are typically subtle, consisting of mouth or face movements or subtle body movements or random walking. Higher-amplitude limb myoclonus or forceful eyelid polymyoclonias (with photosensitive EEG) are not consistent with CAE and suggest JME or Jeavon syndrome, respectively.

TREATMENT STRATEGY

Ethosuximide, lamotrigine, or valproic acid appear to be approximately equally efficacious for CAE. Ethosuximide is essentially limited to absence seizures, whereas lamotrigine and valproic acid cover a much wider range of epilepsies (which may be important if JAE is part of the differential). Each drug also has its own side-effect profile that impacts its selection. Most commonly, ethosuximide may cause GI complaints, sleep disturbances, hiccups and, rarely, rashes; lamotrigine may cause serious rashes, especially if increased too rapidly; and valproic acid may cause weight gain, female hormonal changes, and platelet abnormalities. Occasionally, a clinician will find it necessary to combine these drugs to treat more refractory cases. Intravenous benzodiazepines are effective for absence status epilepticus, and sometimes short-term oral benzodiazepines have a role in CAE management. Ensuring seizure control with an EEG performed once clinical seizure freedom is achieved seems prudent, given the subtle nature of the seizures.

Decisions as to when medication can be stopped are often difficult. A conservative approach is to wait a year or two after diagnosis, or until the child enters puberty, whichever is earlier. Then, check an EEG with HV while on medication. If no 3 Hz bursts are recorded, taper meds. If clinical seizures occur, resume meds and repeat process again in the future. If the child remains seizure free after medication taper, consider checking another EEG to confirm seizure freedom.

LONG-TERM OUTCOME

The outcome of CAE is usually favorable with cessation of seizures and elimination of medications by the end of puberty, if not sooner. Recent work suggests that children with CAE have a greater incidence of behavioral and learning problems than age-matched peers; however, that data is still being collected as part of a National Institutes of Health (NIH)-sponsored multicenter CAE trial (Childhood Absence Epilepsy Rx PK-PD-Pharmacogenetics Study; T. Glauser, principal investigator).

A small percent of children with absence seizures evolve to other seizure types, perhaps indicating that they may have had a very early onset of JAE or JME rather than CAE.

PATHOPHYSIOLOGY/NEUROBIOLOGY OF DISEASE

The pathophysiology of absence seizures has been linked to oscillatory thalamic–cortical potentials, calcium currents, and the interaction of GABAergic neurons. It does seem clear that it differs from the pathophysiology of other epilepsies, which may, in part, explain the unique efficacy of ethosuximide in this syndrome. See suggested references for a more detailed discussion of proposed CAE mechanisms. Multiple genetic regions have been associated with CAE, and no diagnostic clinical genetic tests have yet been developed for this disorder.

CLINICAL PEARLS

1. CAE can often be reliably diagnosed and followed with a simple 3–5 minute office HV trial. Having the child blow a suspended piece of paper (to keep it from dropping) or blow a pinwheel are more effective than nondirected HV.
2. Although exercise-induced HV rarely induces seizures, there may be some risk to HV in preparation for a prolonged underwater swim or a breath-holding contest at the pool.
3. Very short or fragmentary spikes may occur during a sleep EEG in a child with CAE and not be indicative of inadequate seizure control.
4. The automatisms occurring with CAE may appear complex. A premature (and incorrect) diagnosis of partial seizures will often lead to the wrong choice of meds.
5. The nomenclature of absence epilepsies can be confusing: "typical" refers to a typical EEG and typical clinical manifestation of CAE. "Atypical" refers to a more complex clinical picture, often in a neurologically impaired child, often with mixed seizure types and mixed features on EEG.
6. Carbamazepine, oxcarbazepine, gabapentin, tiagabine, pregabalin, and vigabatrin may facilitate both absence and myoclonic seizures. Phenobarbital and phenytoin are simply ineffective for CAE.

SUGGESTED READING

Arzimanoglou A, Guerrini R, Aicardi J. (2004). Epilepsies with typical absence seizures. In *Aicardi's Epilepsy in Children, 3rd ed.*, pp. 88–104. Philadelphia: Lippincott, Williams & Wilkins.

Nordli D. (2005). Idiopathic generalized epilepsies recognized by the international league against epilepsy, *Epilepsia* 46, (s9), pp. 48–56.

Panayiotopoulos CP. (1998). Absence epilepsies. In *Epilepsy: A Comprehensive Textbook*, Ed. Engel J, Pedley TA. Philadelphia: Lippincott–Raven.

23 Panayiotopoulos Syndrome

Korwyn Williams, M.D., Ph.D.

CONTENTS

CASE PRESENTATION

The patient is a right-handed 5-year-old girl without a significant past medical history and with normal development. She was urgently seen in the emergency department for altered mental status. Her parents brought her because of unusual behaviors one morning: repeatedly asking the same question and preparing for school on a Sunday. She woke up that morning complaining of stomach upset and vomited three times. Her mother thought she looked pale and her eyes appeared dilated. After the unusual behavior began, the parents brought her to the hospital. The parents denied fever, ill contacts, or access to medications or household chemicals. Her development was normal, and she had no risk factors for epilepsy. Her family history was unremarkable. On examination, she was mildly tachycardic and appeared pale, but was not ill-appearing or diaphoretic. She answered questions with inappropriate responses and did not cooperate fully with the exam. Her pupils were dilated but reactive.

During the course of the interview, the patient's eyes deviated to the right, and soon her right arm exhibited clonic activity. She was given lorazepam, which aborted the seizure. The patient appeared sleepy afterwards. Computed tomography (CT) of the head was unremarkable. Other studies, including spinal fluid analysis, proved to be normal. An electroencephalogram (EEG) was notable for intermittent slowing over the left central region and independent epileptiform discharges during sleep over the left centrotemporal and occipital regions, but no electrographic seizures were recorded. On further questioning of the parents, they recalled a somewhat similar episode three months prior, where she awoke from sleep complaining of stomach upset. She appeared pale, retched, and vomited. She seemed "out of it" for several minutes, which they attributed to the vomiting. They also recalled that she complained frequently of stomach upset. A diagnosis of autonomic seizures (Panayiotopoulos type?) was entertained. The patient returned to baseline in a few hours. Her exam at that time was unremarkable. She was prescribed rectal diazepam but was not started on daily antiepileptic therapy at the time. Subsequently, she has done well without further confusional episodes in the past 3 years.

DIFFERENTIAL DIAGNOSIS

In this case presentation, the diagnostic possibilities encompass more than just seizures. Given the prominent autonomic findings and confusional state, toxic ingestions and metabolic disturbances need to be quickly excluded. The history of emesis, confusional state, and seizure should raise concerns for a meningoencephalitis. Other less likely considerations include stroke and migraine variants.

The patient's rapid recovery to baseline is reassuring and suggests an unprovoked autonomic seizure. Although autonomic findings can occur in almost any type of seizure, autonomic seizures are those in which autonomic features are either present at onset and predominate and/or are a clinically significant component of the seizure. The most common signs and symptoms are nausea, retching, and vomiting; pallor or flushing; pupillary dilatation; syncope; apnea; and tachycardia. Uncommon signs or symptoms include diarrhea, hippus, erections, or cardiopulmonary arrest. These symptoms and findings can occur in isolation or in any combination.

In the pediatric age group, the most common idiopathic autonomic syndrome is Panayiotopoulos syndrome (PS), or early-onset benign childhood seizures with occipital spikes. This syndrome is thought to account for between 3–6% of all pediatric epilepsy cases (up to 15 years of age). It was initially described in 1989 as nocturnal seizures characterized by tonic eye deviation and vomiting, which would occasionally secondarily generalize. Prominently, occipital epileptiform abnormalities are found on EEG.

Retrospective and prospective studies and a recent consensus statement reviewed the features of PS. The peak onset is between 3–6 years of age (range 1–14 years). The majority of seizures occur during sleep. In a large series by Carballo, the most common ictal finding was pallor (94%), followed by ictal vomiting (82%), nausea (21%), and retching (16%). Eye and/or head deviation occurred in almost 90%, focal clonic activity in 31%, generalized tonic–clonic activity in 36%, and visual symptoms in 10%. Three-quarters of those without a generalized seizure had impaired consciousness with the seizure, and 10% of patients experienced a rolandic seizure, coincident with or after remission of the autonomic seizures.

The seizures are infrequent; approximately 80% of the patients have five or less seizures, with half experiencing only one seizure. A third may experience a seizure during wakefulness, but it is very uncommon not to experience a nocturnal seizure. The seizures themselves tend to be long; a third experienced partial status epilepticus (pallor and/or vomiting, eye deviation, and impairment of consciousness with eventual convulsive activity). With briefer seizures, the average duration is about 9 minutes.

Gastaut syndrome, or late-onset benign childhood seizures with occipital spikes (to be distinguished from the Lennox–Gastaut syndrome), deserves a brief mention in contradistinction to PS. Its peak onset is 7–9 years of age. The seizures usually last only a few minutes and begin with visual hallucinations or amaurosis, eye deviation, and convulsive activity. Unfortunately, they tend to recur, and medical treatment is usually indicated.

DIAGNOSTIC APPROACH

The cornerstone of the evaluation is a sleep-deprived EEG. The most common finding is occipital epileptiform discharges (synchronous or unilateral), which are seen in three-quarters of the patients. However, it should be noted that 10–25% exhibit only extraoccipital epileptiform discharges (typically frontal or temporal). The location of these discharges can vary even in the same patient. Less commonly, the EEG can be normal. Sleep activates the discharges significantly and may be the only time epileptiform discharges are seen. Ictal discharges originate from occipital, frontal, and temporal regions, but seizure manifestations can be subtle (in one case, the only ictal change was tachycardia for 10 minutes before eye deviation and convulsive activity occurred). Autonomic seizures can be due to symptomatic, cryptogenic, and idiopathic etiologies. PS itself is a diagnosis that is most firmly established in hindsight. Therefore, neuroimaging to evaluate for symptomatic etiologies would be prudent.

TREATMENT

Because the seizures in this Panayiotopoulos syndrome are infrequent and uncommon, these patients are usually not placed on antiepileptic drugs. However, if the seizures are frequent or concerning, carbamazepine, valproic acid, and topiramate, among others, have been used. In one large series, almost 90% were seizure free after treatment, but approximately 5% continued to have seizures despite treatment. For those who present with status epilepticus, rectal diazepam is a reasonable therapy to prescribe.

LONG-TERM OUTCOME

The prognosis for this condition is generally favorable. Only one-third of patients will experience a second seizure. The seizures are infrequent and almost always remit within 6 years (most within 3 years) of the first seizure. Notably, the epileptiform abnormalities may persist after seizures have remitted. The overall neurodevelopmental trajectory is normal, as is expected for a benign epilepsy of childhood, but there are only limited studies in this population. A fraction of these children can develop seizures during late childhood and adolescence, typically rolandic seizures or absence seizures, which almost always remit as well.

NEUROBIOLOGY/PATHOPHYSIOLOGY OF DISEASE

The neurobiological correlation between the epileptiform discharges and the semiology is unknown. Autonomic centers typically are midline deep structures of the brain, so the basis of the epileptiform discharges leading to autonomic seizures are unclear. In addition, seizures in the same patient may lead to different autonomic symptoms, suggesting that epileptic propagation can follow different pathways. A small case series has localized various autonomic phenomena (i.e., ictal bradycardia, asystole, ictal pallor) using video-EEG monitoring for temporal lobe seizures. Approximately 30% of these patients have first-degree relatives with seizures, but the genetics of the syndrome are poorly defined. One recent case report of a child with atypical PS (retching, vomiting, hypotonia, loss of consciousness, cardiorespiratory arrest) identified a mutation in a sodium channel gene (*SCN1A*).

CLINICAL PEARLS

1. Panayiotopoulos syndrome is an age-defined benign epilepsy of childhood with peak incidence between 3–6 years of age.
2. The seizures are autonomic in nature, with the most common signs being pallor and recurrent ictal vomiting, but can be variable. Later in the disease course, consciousness may become impaired, eye deviation may be seen, and secondary generalization can occur.
3. The seizures are typically long (at least 5 minutes), and a significant number may present in autonomic status epilepticus.
4. The EEG may show a predominance of occipital epileptiform discharges, but the discharges may be multifocal or even purely extraoccipital.
5. Antiepileptic drug therapy is often not necessary, and the long-term prognosis is very good for eventual seizure remission.

SUGGESTED READING

Caraballo R, Cersosimo R, Fejerman N. (2007). Panayiotopoulos syndrome: A prospective study of 192 patients. *Epilepsia* 48, 1054–1061.
Ferrie CD, Caraballo R, Covanis A. et al. (2006). Panayiotopoulos syndrome: A consensus view. *Dev. Med. Child Neurol.* 48, 236–240.
Ferrie CD, Caraballo R, Covanis A. et al. (2007). Autonomic status epilepticus in Panyiotopoulos syndrome and other childhood and adult epilepsies: A consensus view. *Epilepsia* 48, 1165–1172.

Panayiotopoulos CP. (1989). Benign nocturnal childhood occipital epilepsy: A new syndrome with nocturnal seizures, tonic deviation of the eyes, and vomiting. *J. Child. Neurol.* 4, 43–49.

Tedrus GM, Fonseca LC. (2006). Autonomic seizures and autonomic status epilepticus in early onset benign childhood occipital epilepsy. *Arq. Neuropsiquiatr.* 64, 723–726.

DIFFERENTIAL DIAGNOSIS

Recurrent staring spells are a frequent cause for referral to child neurologists and electroencephalogram (EEG) laboratories. These events may initially be noted by parents or teachers, and may be difficult to differentiate from behavioral inattentiveness—particularly in children with attention deficit disorder. However, some of the children will have epilepsy where the staring spells represent seizures. The two main seizure types that present as staring spells, namely, absence and complex partial, can usually be differentiated based upon their clinical features. Absence seizures tend to last several seconds in duration, manifest as behavioral arrests with brief interruptions in speech or activities, and have no postictal state. Longer absences may be associated with simple motor automatisms. In untreated children, absence seizures may occur hundreds of times per day.

Complex partial seizures also present with staring spells, but their interruption in awareness is less complete. They may be partially responsive, and may follow simple commands. The seizures are more likely to be associated with motor automatisms and are typically followed by confusion and drowsiness in the postictal state. Most complex partial seizures last 1–2 minutes in duration, rarely occurring more than a few times per week or month. Some children may experience a simple partial seizure (aura) before losing awareness with the complex partial seizure. The character of the aura may suggest the region of the brain where the seizure begins. Children may come to medical attention following a secondarily generalized seizure, with tonic or clonic activity that often frightens parents or caregivers. RM is likely experiencing complex partial seizures, and his aura suggests temporal lobe involvement. Because motor automatisms are often ipsilateral to the ictal onset zone, his seizures are probably arising from the right temporal lobe.

DIAGNOSTIC APPROACH

An EEG will quickly differentiate between absence and complex partial seizures. Absence seizures are associated with bursts of generalized 3 Hz spike-and-slow-wave activity (see Chapter 21), and complex partial seizures with focal spike-and-slow-wave discharges. The location of the spikes often suggests the region of the brain from where the seizures are arising. However, some children with epilepsy do not have any spikes on a routine EEG, and not all children with spikes on their EEG will experience seizures or develop epilepsy. If the nature of the spells remains unclear after a thorough history, physical examination, and routine EEG, then a more prolonged EEG/video monitoring study may be indicated. This allows the clinical events to be recorded and time-synchronized EEG to be reviewed. The vast majority of seizures have clear EEG abnormalities during them, and this technique is also the gold standard for localizing where in the brain the seizures are starting. RM's EEG shows right anterior temporal spike and slow–wave discharges (Figure 24.1).

Most children with partial seizures and focal spikes on EEG require the higher spatial resolution of MRI to search for underlying structural abnormalities that may not be seen on CT. These include scars or gliosis from a remote injury, brain tumors, arteriovenous malformations, and abnormalities of cortical development such as

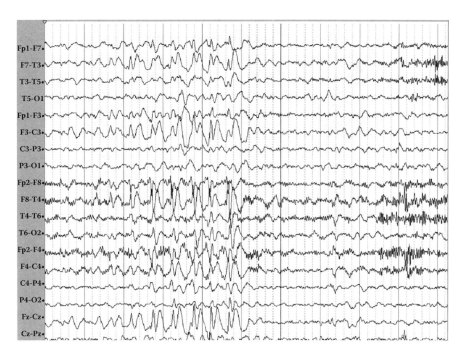

FIGURE 24.1   RM's EEG showing right anterior temporal lobe spike and slow-wave discharges.

focal cortical dysplasias. Other techniques, such as ^{18}F-flourodeoxyglucose positron emission tomography (FDG-PET), may be employed to identify focal brain abnormalities, and may demonstrate an area of hypometabolism. Single-photon emission computed tomography (SPECT) may show interictal hypoperfusion associated with the abnormal area but demonstrate increased perfusion during the seizure (ictal-SPECT). Magnetoencephalography (MEG) allows for colocalization of interictal abnormalities and brain MRI, and may be useful in poorly localized scalp EEG findings. RM has a small focal cortical dysplasia in his right temporal lobe (Figure 24.2). He, therefore, has symptomatic localization-related epilepsy with complex partial seizures, associated with a focal cortical dysplasia in his right temporal lobe. A partial list of etiologies associated with localization-related epilepsy is listed in Table 24.1.

TREATMENT STRATEGY

RM requires intervention due to the frequent recurrent nature of his seizures. The seizures place him at risk for accidents and injuries, may impair cognition and his academic performance, and may have major deleterious psychoemotional effects, particularly when seizures occur at school. Fortunately, there has been an ever-growing list of newer antiepileptic drugs (AEDs) and nonpharmacologic therapies available to practitioners who manage childhood epilepsy. Traditionally, the

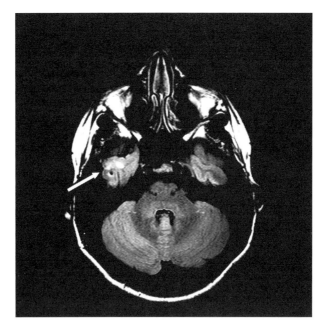

(A)

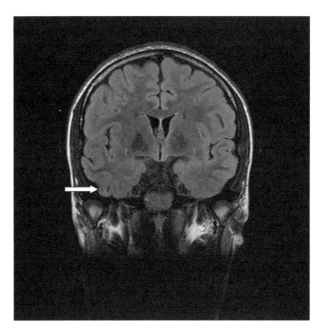

(B)

FIGURE 24.2   RM's MRI scan with axial (A) and coronal (B) images showing a right temporal lobe focal cortical dysplasia.

TABLE 24.1
Etiologies of localization-related epilepsy in children

Idiopathic

Benign childhood epilepsy with centrotemporal spikes (benign rolandic epilepsy)
Benign childhood epilepsy with occipital paroxysms (panayiotopoulos syndrome)

Symptomatic

Neurodevelopmental

Focal cortical dysplasia

Schizencephaly

Periventicular nodular heterotopias

Hemimegalencephaly

Polymicrogyria

Pachygyria

Arteriovenous malformations

Neurocutaneous

Tuberous sclerosis complex

Sturge–Weber syndrome

Neurocutaneous melanosis sequence

Neoplastic

Low-grade developmental tumors are the most common, and include gangliogliomas and dysembryoplastic neuroepithelial tumors (DNET)

Posttraumatic

Focal brain injury following trauma to the brain, particularly if hemorrhagic, and also includes poststroke, postmeningitic, postencephalitic, posthypoxia-ischemia

Metabolic diseases

Many inborn errors of metabolism, such as MELAS, are associated with localization-related epilepsy

Autoimmune

Includes immune-mediated disorders such as Rasmussen's disease and the antiphospholipid antibody syndrome

medications have been separated into "older" and "newer" groups based upon their historic regulatory approval and appearance in the U.S. marketplace. Typically, when a medication is first approved for epilepsy, it receives an "on-label indication" for add-on (adjunctive) therapy for partial-onset seizures in adults. Then, as experience grows and other studies are done, the use of the drug may expand to other seizure types and younger age groups as deemed appropriate. As a broad generalization, most practitioners who specialize in epilepsy (epileptologists) would now prefer to initiate drug therapy with one of the newer medications. Research studies and clinical experience have shown that the newer medications may not be more efficacious than the older drugs, but they do appear to be safer, better tolerated, and have fewer drug-to-drug interactions. The AED chosen for initial therapy should be one that is highly effective for a particular seizure type or epilepsy syndrome, and be safe and well tolerated. Single drug therapy (monotherapy) is the goal of epilepsy treatment as it is associated with better compliance, fewer adverse effects, less potential for teratogenicity during pregnancy, and lower cost than polytherapy. Drug interactions are also avoided, and the pharmacokinetics are simplified. There are several appropriate AED treatment options for RM, and his neurologist chose oxcarbazepine.

LONG-TERM OUTCOME

Once on a low-therapeutic dosage of oxcarbazepine, RM stopped having clinical seizures. However, after approximately 1 year, the seizures recurred and only transiently responded to sequential dosage increases. He was gradually changed to lamotrigine, to which topiramate was added due to persistent seizures. RM was referred to a tertiary epilepsy center where he underwent a presurgical evaluation and was found to have seizures arising from the right temporal lobe. The literature reflects that only 50% of patients will remain seizure free with the first medication chosen, and only 60–70% of patients will have complete seizure control with multiple medication trials. Patients who fail two appropriately selected antiepileptic medications are diagnosed to have medically intractable or medically refractory epilepsy, and non-pharmacologic treatments should be considered. These include resective epilepsy surgery, vagus nerve stimulation, and the ketogenic diet. Whenever possible, resective epilepsy surgery is usually considered first, as it offers the best hope for complete seizure control or remission. The best candidates for resective epilepsy surgery are those with a symptomatic etiology who have a lesion visible on MRI. Patients with temporal lobe lesions overall have the best outcome, particularly those with mesial temporal sclerosis. Seizure-free rates approach 70–80% with lesional temporal lobe epilepsy, and are 50–60% following extratemporal resections. Patients with the lowest probability for complete seizure control are those with nonlesional extratemporal epilepsy, particularly from the frontal lobe. A right anterior temporal lobectomy was performed, and RM tolerated the procedure well and without any adverse effects. He was seizure free postoperatively; topiramate was stopped after 3 months, and lamotrigine after an additional 6 months. He has remained seizure free off medications and has excelled academically and socially.

PATHOPHYSIOLOGY/NEUROBIOLOGY OF DISEASE

A seizure represents the clinical expression of abnormal, excessive, and synchronous discharges of neurons primarily in the cerebral cortex that is usually self-limited, lasting seconds to a few minutes. Seizures are characterized on EEG by sustained abnormal electrical activity that has a relatively discrete beginning and end, and goes through an evolution characterized by changing shapes (morphology) and amplitude (voltage) of the abnormal discharges (usually spikes or rhythmic waves). A focal or partial seizure has a restricted regional onset that may or may not be followed by spread to neighboring or remote brain regions. It may also spread to deep subcortical regions and result in a generalized tonic–clonic seizure. Partial seizures may start in a "silent" area of the brain, such as the frontal lobe, and become clinically apparent only when they spread to neighboring cortex such as the precentral gyrus of the frontal lobe or the hippocampus of the temporal lobe. In these cases, EEG/video monitoring can be critical to the detection of the site of seizure onset.

An individual is considered to have epilepsy when seizures recur spontaneously over a period of time. Epilepsy is not a specific disease but rather a condition arising from a variety of brain disturbances caused by virtually any pathological insult of the cortex. Specific etiologies range from tumors to genetic channelopathies. Brief, unsustained bursts of abnormal neuronal discharges (interictal discharges) occur between the actual seizures, and can be recognized on the EEG. These transient discharges are called "spikes" or epileptiform discharges, and may also be referred to as "potentially epileptogenic" by the electroencephalographer. They may occur in the EEGs of 3 to 5% of normal children, and are more frequent in near relatives of individuals with seizures, particularly in those families with a genetic epilepsy.

Age-related epilepsies that occur in otherwise normal children who have specific EEG spike patterns and are due to genetic ion channel or receptor defects are referred to as "idiopathic." Those that are secondary to some type of physical or metabolic disorder affecting the brain are called "symptomatic" when the cause is known, or "cryptogenic" when the exact cause cannot be identified. Seizures that are the result of a past injury or insult are considered "remote symptomatic."

CLINICAL PEARLS

1. Differentiation of absence and complex partial seizures often relies on EEG characteristics and may require prolonged EEG/ video monitoring in some cases.
2. MRI should be performed in most children with partial epilepsy instead of CT to evaluate for subtle abnormalities, such as focal cortical dysplasias.
3. Most patients with partial epilepsy will respond to treatment with antiepileptic medications. Failure to respond to two or more medications should prompt a referral to an epilepsy specialist, as many of these patients remain refractory to medications.
4. Resective epilepsy surgery offers the best chance of becoming seizure free in appropriately selected refractory patients.

SUGGESTED READING

Andermann, F., Kobayashi, E., and Andermann, E. (2005). Genetic focal epilepsies: state of the art and paths to the future. *Epilepsia* 46(Suppl. 10), 61–67.

Cross, J.H., Jayakar, P., Nordli, D., Delalande, O., Duchowny, M., Wieser, H.G., Guerrini, R., and Mathern, G.W. (2006). Proposed criteria for referral and evaluation of children for epilepsy surgery: recommendations of the Subcommission for Pediatric Epilepsy Surgery. *Epilepsia* 47, 952–9.

Fogarasi, A., Tuxhorn, I., Hegyi, M., and Janszky, J. (2005). Predictive clinical factors for the differential diagnosis of childhood extratemporal seizures. *Epilepsia* 46, 1280–1285.

Hadjiloizou, S.M. and Bourgeois, B.F. (2007). Antiepileptic drug treatment in children. *Expert Review of Neurotherapeutics* 7(2), 179–193.

Keene, D.L. (2006). A systematic review of the use of the ketogenic diet in childhood epilepsy. *Pediatric. Neurol.* 35(1), 1–5.

Malphrus, A.M. and Wilfong, A.A. (2007). Use of the newer antiepileptic drugs in pediatric epilepsies. *Curr. Treat. Options Neurol.* 9(4), 256–267.

Raybaud, C., Shroff, M., Rutka, J.T., and Chuang, S.H. (2006). Imaging surgical epilepsy in children. *Child's Nerv. Syst.* 22(8), 786–809.

Sabaz, M., Lawson, J.A., Cairns, D.R., Duchowny, M.S., Resnick, T.J., Dean, P.M., Bleasel, A.F., and Bye, A.M. (2006). The impact of epilepsy surgery on quality of life in children. *Neurology* 66(4), 557–561.

Sankar, R. (2004). Initial treatment of epilepsy with antiepileptic drugs: pediatrics issues. *Neurology* 63(10 Suppl. 4), 30–39.

Wolf, S.M. and McGoldrick, P.E. (2006). Recognition and management of pediatric seizures. *Pediatr. Ann.* 35(5), 332–344.

25 LENNOX–GASTAUT SYNDROME

Jong M. Rho, M.D.

CONTENTS

CASE PRESENTATION

This is a nearly 3-year-old boy who was completely well until 6 months ago when he experienced two brief (<2 minute) generalized convulsions 1 month apart, both associated with fever. Although these were diagnosed as simple febrile seizures, he was referred to a pediatric neurologist who ordered a routine 1.5T brain magnetic resonance imaging (MRI) scan. This study was interpreted as normal. However, the outpatient electroencephalogram (EEG) study revealed generalized background slowing and generalized atypical spike–wave discharges occurring at a frequency of 2.0–2.5Hz (See Figure 25.1). A follow-up 24-hour video-EEG study captured multiple generalized myoclonic and tonic seizures, both associated with generalized spike–wave discharges, followed by voltage suppression. No focal or lateralizing features were noted. This patient was placed on valproic acid monotherapy. Despite serum levels (in the 100–120 range), he continued to have frequent seizures and then developed "drop" seizures that became progressively more frequent. Further medication trials with topiramate, zonisamide, levetiracetam, and clonazepam were unsuccessful, and he was ultimately placed on the ketogenic diet. His generalized myoclonic seizures improved substantially, but he was still experiencing several drop seizures per day. Notably, his speech and attention became slowly impaired, and he developed difficulty walking, with mild ataxia. A high-resolution (3T) brain MRI was interpreted as normal, and a comprehensive metabolic/genetic workup

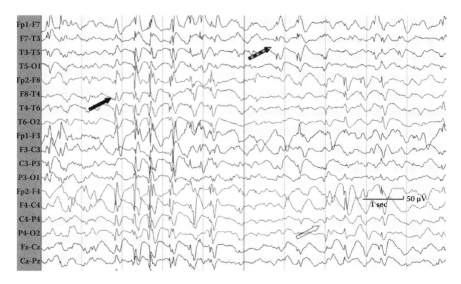

FIGURE 25.1   Generalized 2–2.5 Hz slow spike–wave complexes (solid arrow) and spikes lateralized to the left and right hemispheres (dashed and open arrow, respectively).

(including genetic testing for severe myoclonic epilepsy of infancy) failed to reveal an etiology. As his neurological condition steadily worsened, he underwent an anterior two-thirds corpus callosotomy as a palliative procedure, to which he responded favorably. Although he experienced a transient left foot drop after surgery, his speech returned slowly (but not fully), and within a month he was able to put three words together again. He was weaned off the ketogenic diet, and remained on a regimen of zonisamide and levetiracetam. His gait and ataxia improved, and he was having only 1–2 "drop" seizures per day, but none of the myoclonic or tonic seizures.

DIFFERENTIAL DIAGNOSIS

Lennox–Gastaut syndrome (LGS) first appeared in the medical literature in 1969, even though important clinical and EEG features in this subgroup of epileptic patients had been noted as early as 1939. This syndrome comprises a clinical triad consisting of (1) diffusely slow spike–wave discharges (occurring at a frequency of 1.5–2.5 Hz), (2) psychomotor retardation, and (3) multiple electroclinical seizure types refractory to medical therapy (including generalized tonic, atonic, atypical absence, myoclonic, and tonic–clonic seizures). The hallmark seizure type is the generalized tonic seizure that most often occurs while falling asleep. In most cases, LGS manifests between 2–8 years of age, represents 3–10% of all pediatric epilepsies and, in this condition, affects males more frequently than females. There are two general subtypes: (1) cryptogenic (i.e., there is no identifiable cause) in approximately a third of patients, and (2) symptomatic (i.e., associated with a remote brain injury, usually acquired during the perinatal period or early infancy). It is not uncommon for LGS to be preceded by a history of infantile spasms.

The major differential diagnostic consideration is myoclonic–astatic epilepsy (aka Doose syndrome), which appears between 2–5 years of age, often with generalized tonic–clonic seizures. The affected child, however, is developmentally normal prior to onset of seizures, but developmental decline can occur if they are not controlled adequately. Within several months of onset, the characteristic drop attacks occur, along with atypical absence seizures. There is also the syndrome of continuous spike–wave discharges in slow-wave sleep (aka electrical status epilepticus in slow-wave sleep, or ESES), which can result in speech/language regression, slow spike–wave discharges, as well as atonic and atypical absence seizures. However, generalized tonic seizures do not occur with ESES. Finally, epileptic spasms (i.e., spasms that persist beyond infancy) can also produce drop attacks.

Careful video-EEG monitoring of episodes can help distinguish between LGS and these other rare entities. If the clinician obtains a generic history of drop attacks without other ancillary data (most importantly, the EEG), then other diagnostic considerations would include syncope (neurogenic or cardiogenic), cataplexy (seen in narcoleptic patients), and hyperekplexia (or "startle disease"). It is important to note, however, that there are several epileptic syndromes and epileptic encephalopathies that share some but not all clinical and/or EEG features of LGS, and many LGS patients may lack all of the characteristic features, especially during the early stages of disease evolution.

DIAGNOSTIC APPROACH

There are two general goals of the diagnostic approach toward LGS. The first is to establish a clear diagnosis of the syndrome, and the second is to determine whether the particular patient has the cryptogenic or symptomatic subtype (and with the latter, a specific etiology). The single most important diagnostic tool is the EEG, and as is often the case, long-term video-EEG monitoring to accurately define specific seizure types and specific EEG abnormalities (both ictal and interictal). The next most useful diagnostic test is the brain MRI because a variety of structural lesions can produce LGS; these include destructive pathologies such as meningitis/encephalitis, hypoxic–ischemic injury, stroke, and trauma, as well as developmental anomalies such as tuberous sclerosis and disorders of neuronal migration (e.g., cortical dysplasia). A head computed tomography (CT) scan can be valuable in demonstrating the presence of calcifications, which can be seen in conditions such as congenital cytomegalovirus (CMV) infection. Depending on other features of the physical and neurological examinations, specific metabolic and/or degenerative disorders affecting the central nervous system should be considered, requiring specialized genetic testing, biochemical assays, and biopsies. It should be noted that upwards of 30% of LGS patients have no evidence of preexisting brain damage, despite the fact that approximately one-third of them have a history of prior infantile spasms.

TREATMENT STRATEGY

By and large, the treatment of LGS has been challenging and disappointing. The optimum treatment for LGS remains unclear, and no study to date provides solid evidence of any single drug to be highly effective. However, given the multiplicity of seizure types seen in LGS, and the fact that certain drugs such as carbamazepine and phenytoin can exacerbate generalized spike–wave epilepsies, it is generally recommended that broad-spectrum (i.e., effective against both focal and generalized seizure types) antiepileptic medications be utilized. Valproic acid has traditionally been the drug of choice, as it can provide beneficial effects against all seizure types, but caution is advised when using valproic acid in young patients (under 2 years of age) on polypharmacy (as this substantially increases the risk of hepatotoxicity). Other newer broad-spectrum antiepileptics include topiramate, lamotrigine, and felbamate, which have been shown to be effective in double-blind, placebo-controlled trials. However, felbamate is associated with a high incidence of serious side effects such as aplastic anemia/pancytopenia and hepatic failure. Although there are numerous uncontrolled studies of other medications being effective for patients with LGS (including zonisamide, clonazepam, nitrazepam, vigabatrin, prednisone, IVIG, etc.), the reported benefits have been variable, limited, or short-lived. Nonpharmacological options, such as the ketogenic diet, vagus nerve stimulation, and corpus callosotomy (especially for the atonic seizures), have been shown to be beneficial to varying degrees. Investigational agents, such as clobazam and rufinamide, may ultimately provide long-term benefits for patients. In practice, the majority of children with LGS need combination therapy, with both antiepileptic medications and nonpharmacological options. Each patient should be considered individually, with a careful assessment of potential benefits of the chosen therapy weighed against the risks of adverse effects.

LONG-TERM OUTCOME

Long-term prognosis of LGS is generally poor, as medically intractable seizures persist in over 90% of patients, and recurrent episodes of status epilepticus are not uncommon. Not only do patients experience significant neurocognitive impairment, they are also exposed to a 3–7% risk of mortality (often coming from accidents, and as such, protective helmets are often advised for patients with drop attacks). Studies have demonstrated a steady worsening of IQ, and behavioral and psychiatric symptoms. Without significant improvement from medical and/or surgical treatment strategies, LGS can be appropriately considered a progressive epileptic encephalopathy. Unlike infantile spasms, where the majority (>80%) of patients outgrow their seizures (but often followed later in life by other seizure types), patients with LGS may evolve to a more refractory form of epilepsy, that is, epileptic spasms, or continue to exhibit electroclinical features of LGS.

PATHOPHYSIOLOGY/NEUROBIOLOGY OF DISEASE

Although the mechanisms underlying the electroclinical expression of LGS remain unclear, there is a unique age-dependence to the onset, which invokes disturbances in normal maturational processes. The fact that many diverse etiologies can produce the same clinical syndromes suggests a final common mechanistic pathway that may be activated in genetically susceptible individuals. In terms of pathological substrates, both frontal lobes and subcortical structures such as the thalamus have been implicated.

CLINICAL PEARLS

1. A strong clinical index of suspicion can be made on the basis of a thorough medical history. LGS is an early childhood epileptic encephalopathy consisting of multiple seizure types, especially tonic seizures during sleep.
2. A diagnosis can be made on the basis of long-term video-EEG monitoring.
3. Although LGS patients are defined as treatment resistant, there are a number of effective nonpharmacological options, including the ketogenic diet, vagus nerve stimulation, and corpus callosotomy. Additionally, immunomodulation with IVIG or steroids can be helpful in some patients.
4. Most successful treatment strategies appear to involve combination therapies that may provide complementary benefits.

SUGGESTED READING

Blume WT. (2001) Pathogenesis of Lennox–Gastaut syndrome: considerations and hypotheses. *Epileptic Disord.* 3(4): 183–96.

Glauser TA, Morita DA. Lennox–Gastaut syndrome. www.emedicine.com/neuro.

Markand ON. (2003) Lennox–Gastaut syndrome (childhood epileptic encephalopathy). *J. Clin. Neurophysiol.* 20(6): 426–441.

Nabbout R, Dulac O. Lennox–Gastaut syndrome. www.medlink.com/medlinkcontent.asp.

26 Nonconvulsive Status Epilepticus

James J. Riviello, Jr., M.D.

CONTENTS

CASE PRESENTATION

A 6-year-old girl was admitted for evaluation of increasing seizure activity, consisting of generalized tonic–clonic seizures and staring spells. The seizure onset had been approximately 1 year previously, her first EEG showed generalized spike-and-wave activity, and MRI was normal. She was subsequently treated with phenobarbital, valproic acid, topiramate, and lamotrigine, and despite this, her seizures had continued and were actually increasing in frequency. On the day of hospital admission, she was noted to have frequent generalized tonic–clonic seizures as well as staring spells associated with myoclonic movements, but in between these seizures she remained lethargic. She was then admitted for seizure control. No further generalized tonic–clonic seizures were observed.

DIFFERENTIAL DIAGNOSIS/DIAGNOSTIC APPROACH

This case involves a child with a history of recent generalized tonic–clonic seizures, and now, continuous lethargy and frequent staring spells. Because the duration of this episode is longer than 5 minutes, could this represent a case of SE? Using a semiological classification system, status epilepticus (SE) is divided into convulsive SE (CSE) and nonconvulsive SE (NCSE). NCSE is defined as altered awareness associated with electrographic seizure activity, and may occur in either a generalized or focal epilepsy. Although clear tonic or clonic activity is not seen,

CLINICAL PEARLS

1. NCSE may develop after the control of the convulsive movements in CSE.
2. In patients with unexplained alteration of awareness, especially in the PICU or emergency department, NCSE should be considered and an EEG performed. A preexisting history of epilepsy increases the likelihood of NCSE or NCS.
3. Treatment for NCSE typically begins with lorazepam. Intravenous VPA or levetiracetam can also be considered as abortive treatment for absence SE.
4. The outcome of absence SE or autonomic SE is generally favorable, whereas the outcome of NCSE is related to the underlying cause, rather than the NCSE itself.
5. Autonomic SE is likely a focal disorder, and ictal EEG may have rhythmic delta or theta activity with admixed spikes, rather than overt spike-and-wave or focal spike discharges.
6. We recommend that the child with epilepsy carry some sort of identification so that if altered awareness occurs without an available historian, medical providers are made aware of the child's condition.

REFERENCES

DeLorenzo RJ, Waterhouse EJ, Towne AR, Boggs JG, Ko D, DeLorenzo GA, Brown A, Garnett L. (1998) Persistent nonconvulsive status epilepticus after the control of generalized convulsive status epilepticus. *Epilepsia* 39: 833–840.

Panayiotopoulos CP. (2004) Autonomic seizures and autonomic status epilepticus peculiar to childhood: diagnosis and management. *Epilepsy Behav.* 5: 286–95.

Privitera MD, Strawsurg RH. (1994) Electroencephalographic monitoring in the emergency department. *Emergency Med. Clin. NA* 12: 1089–1101.

Tay SKH, Hirsch LJ, Leary L, Jette N, Wittman J, Akman CI. (2006) Nonconvulsive status epilepticus in children: Clinical and EEG characteristics. *Epilepsia* 47: 1504–1509.

Wakai S, Ito N, Sueoka H, Kawamoto Y, Hayasaka H, Tsutsumi H, Chiba S. (1995) Complex partial status epilepticus in childhood. *Pediatr. Neurol.* 13: 137–41.

Wheless JW, Clarke DF, Carpenter D. (2005) Treatment of pediatric epilepsy: expert opinion, 2005. *J. Child. Neurol.* 20(Suppl 1): S1–S56.

Wheless JW, Clarke DF, Arzmanoglu A, Carpenter D. (2007) Treatment of pediatric epilepsy: European expert opinion. *Epileptic Disord.* 2007. 9: 353–412.

27 Focal Cortical Dysplasia

Susan Koh, M.D.

CONTENTS

CASE PRESENTATION

The patient, a 6-year-old right-handed girl, presents with a history of seizures beginning 18 months earlier. The typical episode begins with staring, followed by head turning to the right and left arm extension, occurring 6–10 times daily (usually during sleep), and lasting for 20–30 seconds in clusters. The seizures are not associated with postictal lethargy, and she complains of a vague aura if she is awake. She had been treated with several medications in the past, including zonisamide, carbamazepine, and lamotrigine, none of which she responded to. Her initial brain MRI was interpreted as normal, and the initial electroencephalogram (EEG) was notable for right frontal, central, and temporal spike discharges. There were no perinatal problems, and early developmental milestones were normal. However, once her seizures started, she experienced problems with language and behavior. Her physical and neurologic examinations are normal.

Because of her medical intractability, she underwent an epilepsy surgery evaluation consisting of long-term video-EEG monitoring and additional neuroimaging studies, including a positron emission tomography (PET) scan and another magnetic resonance imaging (MRI) scan combined with MRI-PET fusion. EEG revealed interictal right frontal central and temporal spike discharges with a broad ictal onset over the right frontal temporal region. The MRI showed a possible right superior middle frontal cortical dysplasia. The PET demonstrated hypometabolism over the right frontal region. A magnetoencephalogram (MEG) was performed that showed dipoles over the right superior, middle, and inferior frontal gyrus (see Figure 27.1).

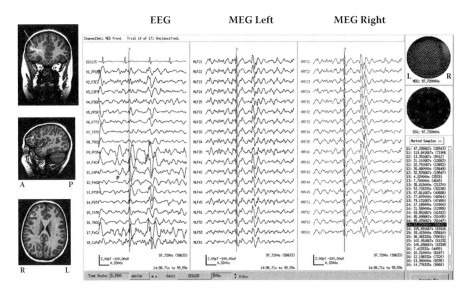

FIGURE 27.1   MEG imaging of this patient superimposed on a T2-weighted MRI, which shows the dipoles located over the right frontal middle gyrus over a subtle cortical dysplasia.

The patient underwent a frontal resection with electrocorticography (ECoG). She has remained seizure free for the last 2 years on one anticonvulsant, and her parents state she is doing well in school and that she is more alert and attentive. Her pathology from the right frontal lobe was reported as mild cortical dysplasia, type 1a, with mild increase in white matter neurons comparable to scattered heterotopic neurons.

DIFFERENTIAL DIAGNOSIS

Etiologies that are seen with intractable complex partial seizures include encephalitis, stroke, tumor, a nonepileptic event (e.g., psychogenic, movement disorder, or parasomnia), and genetic/idiopathic, metabolic disorders that result in encephalopathic epilepsies such as progressive myoclonic epilepsies and malformation of cortical development. The most common malformation of cortical development is a focal cortical dysplasia (FCD). A stroke or tumor is unlikely with a normal neurological examination, although low-grade tumors and developmental neoplasms (such as a dysembryonic neuroepithelial tumor) can occur in the context of a normal exam. In this patient, an infectious etiology is also doubtful because there is no history of an illness, fever, altered mental status, or sick contact. Nonepileptic events due to a psychogenic cause are less probable because the patient has nightly stereotyped episodes. A movement disorder is also implausible in light of a normal past medical history, occurrence during sleep, an aura prior to the event, and altered consciousness postictally. Parasomnias are unlikely because the patient has history of the events occurring during the day as well. Progressive epilepsies are also doubtful because the seizures are partial seizures, and the patient did not experience rapid neurological decline. A benign idiopathic disorder is low on the differential because the seizures are intractable to medications. On the basis of history and physical examination, an FCD would be the best choice, although a low-grade tumor or developmental neoplasm cannot be ruled out.

FCD was first described by Taylor in 1971; it has an estimated prevalence of 5–25% in all epilepsy patients and occurs more commonly in children. Because of advances in neuroimaging, FCD is increasingly recognized as a pathological substrate for medically refractory epilepsy (approximately 76% of patients with FCD do not respond to antiepileptic drugs [AEDs]). In one major pediatric surgical center, 80% of children under 3 years of age having undergone epilepsy surgery were found to have FCD on pathology. In adults, temporal and central lesions are most frequent, whereas parietooccipital lesions are rare. However, in children, frontal and parietooccipital lesions are most frequent. Patients who present early in life, especially before 2 years of age, will develop cognitive impairment; developmental delay is seen in 70–80% of patients. The size of the lesion, the location, and the histopathological subtype influence whether a patient will become mentally disabled. Normal intellect is more frequently found in patients who have a circumscribed FCD, whereas mental retardation is seen in patients who have an early onset of seizures correlated with posterior localized or multilobar lesions.

Multiple seizure types have been described with FCD, most commonly partial seizures, and at times epilepsia partialis continua if the FCD is located in the motor cortex. FCD can also be seen in patients with Lennox–Gastaut syndrome, Landau–Kleffner syndrome, Ohtahara syndrome, and infantile spasms (although usually there are partial seizures intermixed with the spasms). A history of febrile seizures is present in 5.5–25% of all patients with FCD. One-third of patients who have extrahippocampal temporal FCD may have simultaneous hippocampal sclerosis (HS), which is defined as dual pathology. Children with dual pathology often present with febrile seizures and have worse HS compared to others without a febrile seizure history.

DIAGNOSTIC APPROACH

MRI is the imaging modality of choice, having a sensitivity of 63–98%. MRI features of FCD usually include (1) focal cortical thickening with cortical hyperintensity, (2) blurring of the gray-white junction, and (3) signal changes in the underlying white matter where there is hyperintensity on T2-weighted imaging and occasional hypointensity on T1-weighted images. However, 50% of FCDs may not be apparent on MRI. This is especially true in infants because poor myelination makes it difficult to differentiate between gray and white matter, and the lesion may be too small to detect. Therefore, many epileptologists suggest repeating an MRI scan again in an infant after 2 years of age.

Because many patients with FCD have normal MRI findings, functional neuroimaging studies such as PET and single positron emission computerized tomography (SPECT) are useful in determining the epileptogenic area associated with FCD. Most epilepsy centers utilize the [¹⁸F]fluorodeoxyglucose (FDG-PET) tracer to identify abnormal areas of hypometabolism associated with epileptogenic lesions. SPECT is a nuclear-medicine study in which hexylmethylprophylene amineoxine (HMPAO) is injected at the beginning of a seizure to identify an area of increased blood flow. The ictal onset zone may have increased uptake compared to other areas of the brain, aiding in localization. Both SPECT and PET have better resolution when superimposed onto an MRI scan; if utilizing PET, this is called PET-MRI fusion (see Figure 27.2), whereas with SPECT, this is called subtraction ictal SPECT coregistered with MRI (SISCOM). This procedure takes the interictal and ictal SPECTS, subtracts them, and then superimposes them on an MRI scan for better definition. Although available only in limited centers, MEG is a special type of scan that detects and measures magnetic fields generated by electrical activity using superconductive quantum interference devices (SQUIDs). Magnetic fields form dipoles that are parallel to the cortical surface, and are picked up by the SQUIDs. Unlike EEG, MEG is unaffected by overlying tissue or bone that would attenuate electrical potentials between cortex and scalp. Therefore, it is helpful as another evaluation tool when the EEG ictal onset is too broad or unclear. In addition, MEG is beneficial in comparing the epileptogenic zone with the motor, sensory, auditory, visual, or language areas because it utilizes evoked potentials in order to locate these areas.

TREATMENT STRATEGY

As noted earlier, most seizures from FCD are intractable to traditional AED therapy. Medication treatment choice should reflect the underlying epilepsy, and there are no specific medications used for FCD. Multiple drug resistance (MDR) proteins that function to export drugs outside the central nervous system have been found in resected FCD tissue, suggesting a possible mechanism for intractability. Epilepsy surgery is often the most effective treatment for FCD, and a surgical evaluation

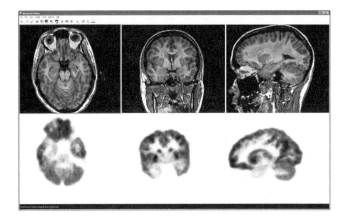

FIGURE 27.2   PET-MRI fusion of a patient with dual pathology consisting of right temporal cortical dysplasia and right hippocampal sclerosis. There is lighter coloration noted over the right temporal lobe in the PET-MRI fusion that corresponds to an area of hypometabolism on PET.

should be undertaken if a patient has failed at least two antiepileptic drugs. However, for patients in whom epilepsy surgery is not feasible, other nonpharmacological options such as the ketogenic diet, vagus nerve stimulation, or transcranial magnet stimulation may be helpful.

LONG-TERM OUTCOME

The majority of studies state that 50–65% of patients with FCD can become seizure free after surgery, and seizure outcome is related to completeness of resection. The area surrounding the FCD may be epileptogenic owing to microscopic spread of CD outside of the MRI lesion, suggesting that a lesionectomy may not suffice. This is especially true for infants because the MRI may underestimate the borders of the FCD. For adults, studies have demonstrated better seizure-free rates compared to children. This may be because adults exhibit more temporal rather than extratemporal lobe seizures with FCD, and are burdened with a shorter duration of epilepsy, less generalized tonic–clonic seizures, and more focal interictal epileptiform discharges than children. Unfortunately, late recurrence of seizures after several years of seizure freedom can occur after surgery. As with other types of symptomatic epilepsy, development and scholastic achievement are often impaired, but may improve with sustained seizure freedom from epilepsy surgery. In general, patients with a younger age of seizure onset and diffuse dysplastic lesions had the poorest intellectual functioning.

PATHOPHYSIOLOGY/NEUROBIOLOGY OF DISEASE

Histopathology in FCD is often based on Palmini's classification system. Type 1a FCD demonstrates isolated architectural abnormalities, whereas type 1b FCD exhibits additional immature or giant neurons. Type II or Taylor type dysplasia is notable for architectural abnormalities along with dysmorphic neurons or balloon cells, which are unusually large cells that are opalescent with eccentric nuclei. It is unclear what type of cell line constitutes a balloon cell. Type 2a shows additional dysmorphic neurons, and type 2b has balloon cells present. Although balloon cells are often associated with FCD, they can also be seen in tuberous sclerosis.

Types 1a and 1b are related to less severe and later-onset epilepsy, usually localized to the temporal lobe. Types 2a and 2b are more associated with a severe epilepsy syndrome and early-onset seizures involving the frontal lobe, along with poor development. Type 1a is often found in patients with dual pathology. MRI findings with type II are more associated with focal cortical thickening, blurring of the gray-white matter, and hyperintensity of the subcortical white matter on T2-weighted imaging, mostly in extratemporal regions. Type 2b FCD exhibits more fluid-attenuated inversion recovery (FLAIR) signal abnormalities. Type I on MRI demonstrates focal brain hypoplasia with shrinkage and moderate signal intensity alterations in the white matter. At present, there are no consistent relationships between histopathologic classification and seizure-free rates.

The pathogenesis of FCD remains unknown, but is likely multifactorial (and inclusive of genetic influences). One theory suggests that balloon cells are similar to immature or stem cells that arise from postnatal neurogenesis in response to seizures or a type of focal insult. This two-hit theory may explain why perinatal adverse events and head trauma are sometimes associated with FCD. In cases of dual pathology, it is unclear whether the HS is a consequence of increased susceptibility to hyperthermia-induced seizures (as a result of FCD in the temporal lobe), or if the FCD and HS come from a common pathologic mechanism during embryogenesis.

CLINICAL PEARLS

1. FCD is the most common type of malformation of cortical development and is often associated with intractable partial seizures.
2. Dual pathology occurs more frequently in older children and adults when there is a strong history of febrile seizures, followed by a honeymoon period, progressing to intractable and frequent seizures.
3. Nearly 50% of FCD are not visible on conventional MRI. However, other neuroimaging techniques, such as PET, SPECT, and MEG, may provide useful localizing information.
4. Frequently, medications are not effective in FCD, and epilepsy surgery provides the best chance of seizure freedom and improved development if the entire area can be resected.

SUGGESTED READING

Bast T, Ramantani G, Seitz A, Rating D. (2006). Focal cortical dysplasia: prevalence, clinical presentation and epilepsy in children and adults. *Acta Neurol. Scand.* 113, 72–81.

Cepeda C, Andre VM, Levine MS et al. (2006). Epileptogenesis in pediatric cortical dysplasia: the dysmature cerebral developmental hypothesis. *Epilepsy Behav.* 9, 219–235.

Colombo N, Tassi L, Galli C et al. (2003). Focal cortical dysplasia: MR Imaging, histopathologic and clinical correlation in surgically treated patients with epilepsy. *Am. J. Neuroradiol.* 24, 724–733.

Frauser S, Huppertz HJ, Bast T. et al. (2006). Clinical characteristics in focal cortical dysplasia: a retrospective evaluation in a series of 120 patients. *Brain* 12, 1907–1916.

Klein B, Levin BD, Duchowny MS, Lisbre MM. (2000). Cognitive outcome of children with epilepsy and malformations of cortical development. *Neurology* 25, 230–235.

Lawson JA, Birchansky S, Pacheco E. et al. (2005). Distinct clinicopathological subtypes of cortical dysplasia of Taylor. *Neurology* 64(1), 55–61.

Lortie A, Plouin P, Chiron C, Delalande O, Dulac O. (2002). Characteristics of epilepsy in focal cortical dysplasia in infancy. *Epilepsy Res.* 51, 133–145.

Otsubo H, Iida K, Oishi M. et al. (2005). Neurophysiologic findings of neuronal migration disorders: intrinsic epileptogenicity of focal cortical dysplasia on electroencephalography, electrocorticography and magnetoencephalography. *J. Child. Neurol.* 20(4), 357–363.

Palmini A, Najm I, Avanzini G. et al. (2004). Terminology and classification of the cortical dysplasia. *Neurology* 62(6 Suppl. 3), S2–8.

Tassi L, Colombo N, Garbelli R. et al. (2002). Focal cortical dysplasia: neuropathologic sub-types, EEG, neuroimaging and surgical outcome. *Brain* 125, 1719–1732.

Taylor DC, Falconer MA, Bruton CJ, Corsellis JA. (1971). Focal dysplasia of the cerebral cortex in epilepsy. *J. Neurol. Neurosurg. Psychiatry* 34, 369–387.

28 Landau–Kleffner Syndrome

John F. Kerrigan, M.D.

CONTENTS

CASE PRESENTATION

This 8-year-old right-handed girl was initially seen at age 7.5 years at a rural satellite clinic. She had previously been diagnosed with Benign Epilepsy with Centrotemporal Spikes (BECTS) and had complete seizure control on carbamazepine. However, the pediatrician also noted a history of "developmental delay." Prior workup had included a normal brain magnetic resonance imaging (MRI), and an electroencephalogram (EEG) report was available, showing frequent bilateral spikes over the centrotemporal regions, activated during sleep.

The history from her mother indicated that she had normal speech until 3 or 4 years of age, and then lost her language skills. The first sign was that she started speaking "gibberish." Her language skills had stabilized at 5 years of age, with limited improvement since. Her first seizure was at age 4, and she has had a total of eight seizures in her life, with none over the past 2 years. She was initially treated with phenobarbital, then carbamazepine.

On examination, she was a well-appearing and attentive girl. She was well socialized, and eager to please. It was quickly apparent, however, that she had profound language problems. She was unable to follow even one-step verbal commands, but she was very observant to any physical cues to indicate what the examiner requested, and she would do her best to comply. When asked what day it was, she responded with "seven." Her mother indicated that she was guessing that the examiner had asked her to provide her age. She responded with one-word answers at best, many of which were jargon. Her neurological examination, including head circumference, was otherwise normal.

Audiology showed normal hearing by threshold testing. A speech pathology evaluation found her to have profound deficits in both receptive and expressive language. She scored at the 2 year and 7 month age equivalency level on the Receptive One-Word Picture Vocabulary Test, and 2 years 10 months on the Expressive One-Word Picture Vocabulary Test. Her articulation was normal.

A repeat brain MRI was normal. A positron emission tomography (PET) study with injection of ^{18}F-flourodeoxyglucose (FDG) showed bilateral temporal lobe hypometabolism (Figure 28.1). An EEG during FDG uptake showed infrequent spikes over the right central and midtemporal electrodes. She underwent 72 hours of inpatient video-EEG recordings, without seizure events. Her background EEG showed minimal slowing, and frequent right temporal spikes, more abundant with sleep. She did not, however, have continuous spike–waves during sleep. A trial of levetiracetam resulted in no clinical change. She was then treated with intravenous methylprednisolone, 20 mg/kg/day, for 3 days, followed by prednisone 40 mg per day for 3 months, also without clinical change. Prednisone was then tapered and discontinued. Multiple subpial transection (MST) was discussed as a treatment option, but her family did not wish to consider this further.

CLINICAL FEATURES

Described in the landmark paper by William Landau (an adult neurologist) and Frank Kleffner (an audiologist) in 1957, Landau–Kleffner syndrome (LKS) remains a rare but distinctive epilepsy syndrome of childhood. Its prevalence within the population is unknown; however, it accounts for roughly 0.2% of patients attending a pediatric epilepsy clinic. Despite the passage of 50 years, we lack specific diagnostic markers as well as an understanding of the basic cellular and molecular mechanisms that are responsible for this condition.

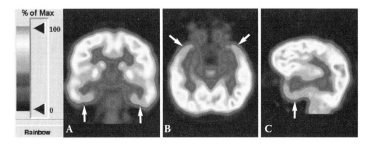

FIGURE 28.1   Positron emission tomography (PET) imaging following intravenous injection with ^{18}F-flourodeoxyglucose (FDG) with representative slices from coronal (A), axial (B), and sagittal (C) planes. Note bilateral temporal lobe hypometabolism, indicated by arrows. PET imaging courtesy of Banner PET Imaging Center, Phoenix, Arizona.

In its purest form, LKS is aptly characterized by its alternative descriptive name, acquired epileptic aphasia. Most commonly, the child has normal speech and language development until the onset of aphasic features between the age of 3 and 8 years. The pace of language deterioration is highly variable, with some children demonstrating slowly progressive changes, whereas others may experience a rapidly progressive or stepwise course over a period of just a few weeks. Classically, the aphasic features consist of a verbal auditory agnosia, in which decoding of verbal stimuli is most severely impaired. However, there is great variability, with some patients completely aphasic and mute in the most severe forms of the syndrome. There may be coexisting behavioral problems, such as attention and mood disturbances. Impairment of socialization may lead to a diagnosis of autism in some patients.

The clinical profile of seizures with LKS is also highly variable. Approximately 20% of LKS patients may have no history of clinically apparent seizures. As a rule, the severity of the seizure disorder is mild, at least when held against the disability caused by the aphasic features. Multiple seizure types are possible in the LKS population, although each individual patient tends to experience stereotyped events. They usually consist of complex partial seizures; however, secondarily generalized tonic–clonic seizures can occur. Seizures are often controlled with antiepileptic drug (AED) management.

EEG studies will usually be abnormal, most commonly with epileptiform spikes or sharp waves that occur over the centrotemporal or midtemporal regions, and usually with bilaterally independent occurrence. Curiously, consistent localization of spike activity to the left (usually language-dominant) hemisphere is not seen. Other patterns also occur, including multifocal independent spikes and generalized spike–wave discharges.

DIFFERENTIAL DIAGNOSIS

There is increasing recognition that a subset of LKS patients will have continuous or nearly continuous spike–wave discharges, usually diffuse or generalized, particularly during sleep. This EEG pattern is associated with an electroclinical syndrome of its own, continuous spike–waves during slow-wave sleep (CSWS), also known as electrical status epilepticus during slow-wave sleep (ESES), which is described separately in Chapter 29. However, the overlap of the clinical features and the lack of clear or specific diagnostic boundaries need to be explicitly acknowledged here. As with LKS, the pathogenetic mechanism (or mechanisms) behind CSWS is unknown. It has been proposed that LKS is entirely a localization-specific subset within CSWS. However, not all children with LKS are shown to have CSWS (although sampling issues may account for some of this). Clearly, overnight EEG studies are more likely to capture CSWS compared to routine outpatient studies.

There is also a lesser degree of overlap with another, more common, epilepsy syndrome of childhood, BECTS (see Chapter 21). BECTS is characterized most commonly by simple partial seizures with orofacial motor features, often easy to control, and associated with surprisingly frequent, high-amplitude spike and aftercoming slow-wave complexes from the central (Rolandic) regions bilaterally. The vast majority of these patients have normal higher cortical function, including language. However, a small number may develop aphasic or dysphasic features, and may be found to have abundant spikes during sleep, or even CSWS. The typical clinical phenotype of BECTS is easily distinguished from LKS, but patients with features common to both syndromes do exist (and may be even more common than the classic LKS phenotype).

As a broad generalization, North American authors have tended to think of these three epilepsy syndromes of childhood as distinct disorders, and they do indeed have separate designations within the 1989 classification of epilepsy syndromes proposed by the International League Against Epilepsy. European authors have more readily embraced the concept that there is a clinical continuum between them. The point to be made here is that patients with overlapping clinical and EEG features will be encountered in routine clinical practice (Figure 28.2).

DIAGNOSTIC APPROACH

Structural imaging studies, including high-resolution MRI, are normal by standard visual analysis. However, a recent report utilizing volumetric analysis identified reduced neocortical volumes bilaterally in four patients with LKS, specifically affecting the superior temporal gyrus and planum temporale, in comparison to a control group of children with frontal or parietal epilepsy. Functional imaging studies have been more revealing, though inconsistent, with decreases in metabolism in the temporal regions by FDG-PET. Magnetoencephalography (MEG) localizes spike dipoles to the perisylvian or insular regions in some cases. Auditory agnosia and receptive language impairment, associated with the language-dominant hemisphere, is the most likely apparent clinical symptom, whereas it is likely the same

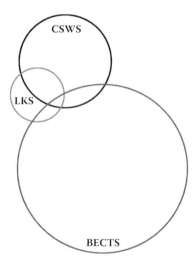

FIGURE 28.2   Venn diagram indicating the overlap of the clinical and EEG features of three pediatric epilepsy syndromes, Landau–Kleffner syndrome (LKS), continuous spike–wave during slow wave sleep (CSWS), and benign epilepsy with centrotemporal spikes (BECTS). The size of each circle represents the relative prevalence of each syndrome.

pathological process occurring simultaneously in the nondominant hemisphere remains clinically less obvious.

It is important to recognize that LKS (or, a syndrome very much like LKS) can occur secondary to other cerebral diseases. These include neurocysticercosis, focal cortical dysplasia, vasculitis, and presumably other focal pathologies that result in cortical injury and epileptic seizures in the susceptible perisylvian regions.

LONG-TERM OUTCOME

The long-term natural history for LKS is generally such that patients stop deteriorating (plateau) after an interval with active regression, usually within 2–4 years of onset. After the syndrome has "run its course," most patients improve with complete disappearance of seizures and improved language capabilities. A complete resolution of the aphasic features is not expected, however, and roughly two-thirds of patients will remain significantly disabled on the basis of language impairment. Patients with earlier onset of symptoms, and those with a more prolonged active phase, have the poorest long-term prognosis.

TREATMENT STRATEGY

Seizures associated with LKS are treated with standard AED therapy. There are no data to demonstrate the superiority of one AED over another, or, indeed, any controlled AED trials for LKS. It must be recognized that some AEDs, among them carbamazepine, phenytoin, and tiagabine, may worsen EEG abnormalities in LKS by unmasking or enhancing generalized spike–wave changes. Accordingly, patients with seizures associated with LKS, particularly those with generalized spike–wave features (with or without CSWS) and generalized seizure types, are probably best treated with "broad-spectrum" AEDs, of which valproic acid (VPA) is the traditional prototype. Benzodiazepines can be effective, and are sometimes recommended in combination with VPA, but their use is limited by sedative effects and by pharmacological tolerance. As noted earlier, the clinical seizure events are usually relatively easy to control. However, stabilization or improvement in the language disturbance often does not parallel seizure control.

An unresolved therapeutic dilemma exists with respect to AED treatment and interictal EEG discharges in LKS. LKS is one of the epileptic syndromes of childhood that is grouped into the epileptic encephalopathies, a broad term for epilepsies that include cognitive and behavioral deterioration, along with seizures. In LKS, this functional deterioration is maximal in the superior temporal neocortex, the cortical region subserving receptive language function. The reason for cognitive and behavioral deterioration with epilepsy are almost always due to multiple factors, including the nonepileptic consequences of the underlying etiology (usually the single most important factor), side effects of AED therapy, the adverse impact of repeated seizures, and the psychosocial consequences of epilepsy. However, there is an emerging concern, perhaps best exemplified by LKS among all epileptic encephalopathies, that the interictal EEG discharges may themselves be implicated as an active causative agent of neuronal injury and dysfunction. This challenges the traditional patient-care maxim of "treat the child, not the EEG." Unfortunately, we do not have proven therapeutic strategies for improving the interictal EEG (i.e., decreasing the abundance of interictal EEG discharges). It is not known if AED therapy can decrease the number of these interictal discharges, and, if suppressed, whether this makes a difference in patient outcome. This is a critically important area of research for LKS and other epileptic encephalopathies of childhood.

Corticosteroids are used for LKS, and several small open-label treatment series have been reported. Although recommended by some authors as a first-line treatment modality for LKS, it appears that most specialists would use corticosteroids if seizures and/or abundant spike–wave patterns fail to respond to AED therapy. There are no controlled trials, and it appears that no two series used the same corticosteroid-dosing regimen. A small number of LKS patients have been reported, following treatment with intravenous immunoglobulin, with favorable results.

MST, a technique in which the intracortical horizontal neuronal processes are transected, has been advocated for treatment-resistant LKS. This strategy is used to surgically treat regions of eloquent cortex that cannot be resected without the likelihood of causing significant new neurological impairment. In practice, and as published in uncontrolled studies, it does appear that MST helps to reduce seizure frequency and halt or even reverse the severity of language impairment, with a good track record for minimizing new neurological deficits. For patients with LKS, MST is applied to the superior temporal gyrus and nearby regions of the cortex in the immediate perisylvian area. MST is usually guided by intraoperative electrocorticography, and appears to be most appropriate in LKS patients with CSWS or very abundant interictal spike–wave activity. However, in individual patients, identifying *exactly* what area of the cortex to treat with MST remains a practical challenge. Additionally, the role of MST relative to corticosteroid therapy remains undetermined.

CLINICAL PEARLS

1. LKS is seen in patients with normal speech and language development until the onset of aphasic features between the age of 3 and 8 years, often associated with recurrent seizures.
2. The EEG typically demonstrates recurrent epileptiform spikes or sharp waves that occur over the centrotemporal or midtemporal regions.
3. Treatment with antiepileptic medications is the mainstay of therapy, whereas corticosteroid therapy and MST may provide benefit in refractory cases.
4. The aphasia of LKS typically plateaus in 2–3 years, and about two-thirds of patients will continue to experience significant language impairment.

SUGGESTED READING

da Silva EA, Chugani DC, Muzik O, Chugani HT. (1997) Landau–Kleffner syndrome: metabolic abnormalities in temporal lobe are a common feature. *J. Child Neurol.* 12: 489–495.

Landau WM, Kleffner FR. (1957) Syndrome of acquired aphasia with convulsive disorder in children. *Neurology* 7: 523–530.

Massa R, de Saint-Martin A, Hirsch E et al. (2000) Landau–Kleffner syndrome: sleep EEG characteristics at onset. *Clin. Neurophysiol.* 111(Suppl. 2): S87–S93.

Morrell F, Whisler WW, Smith MC et al. (1995) Landau–Kleffner syndrome: treatment with subpial intracortical transaction. *Brain* 118: 1529–1546.

Sinclair DB, Snyder TJ. (2005) Corticosteroids for the treatment of Landau–Kleffner syndrome and continuous spike-wave discharge during sleep. *Pediatr. Neurol.* 32: 300–306.

Tsuru T, Mori M, Mizuguchi M, Momoi MY. (2000) Effects of high-dose intravenous corticosteroid therapy in Landau-Kleffner syndrome. *Pediatr. Neurol.* 22: 145–147.

29 Continuous Spike-and-Wave Activity during Slow-Wave Sleep

Kevin Chapman, M.D.

CONTENTS

CASE PRESENTATION

The patient is an 8-year-old male who presents with a 3 year history of seizures. His seizures initially consisted of head deviation to the left with left-sided clonic activity lasting less than 2 minutes in duration. His initial awake EEG demonstrated frequent spikes and sharp waves arising independently from the left frontal and right central head regions. A noncontrast brain magnetic resonance imaging (MRI) was normal. The patient was started on oxcarbazepine, but he continued to have brief seizures averaging one episode per month. During this time, family members as well as teachers noted worsening school performance and increasing irritability. He was switched from oxcarbazepine to valproic acid (VPA), which led to mild improvement in seizure control but not his school performance and neurobehavioral problems. A repeat EEG included a brief period of sleep, and he was noted to have a significant increase in spike activity. An overnight EEG demonstrated nearly continuous high-amplitude spike–wave activity during sleep, with only occasional spikes while awake (see Figure 29.1). The clinical impression was that of continuous spike-and-wave activity during slow-wave sleep (CSWS), and the patient was treated with oral prednisone in addition to VPA. On this regimen, his schoolteachers and

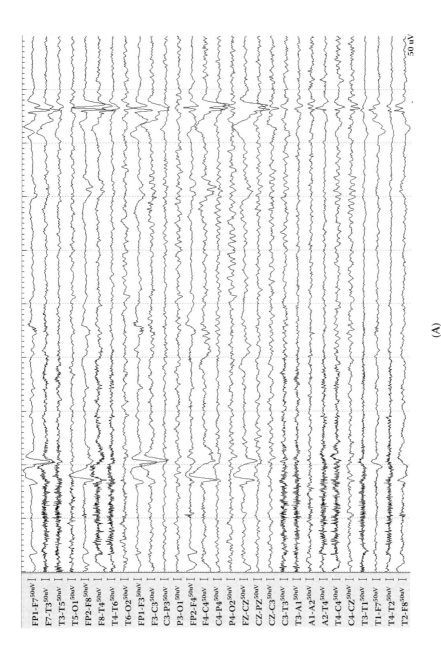

(A)

FIGURE 29.1 (A) Electroencephalogram (EEG) recorded during the awake state, demonstrating a single generalized spike discharge. (B) EEG recorded during sleep, demonstrating nearly continuous generalized 1.5–2 Hz spike–wave activity.

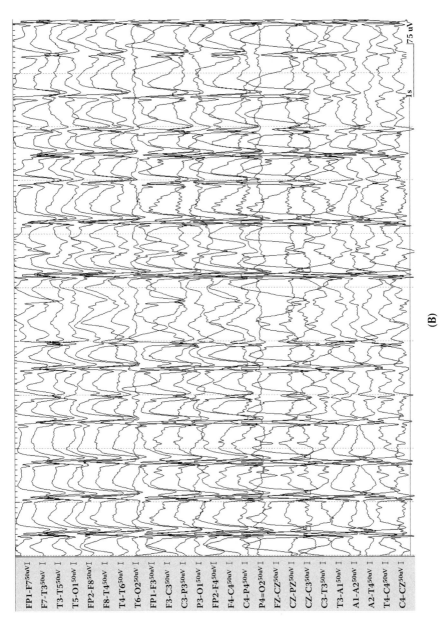

(B)

FIGURE 29.1 (continued)

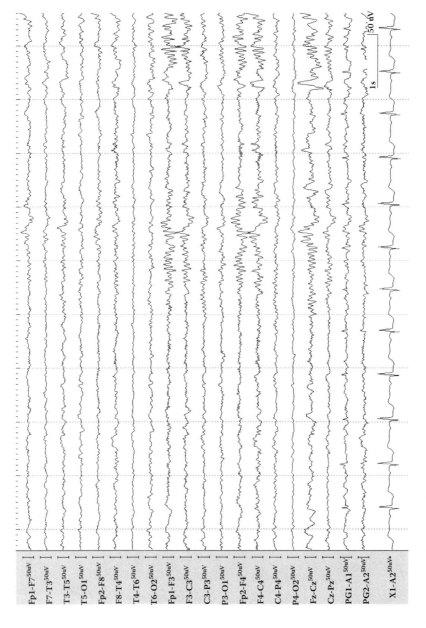

FIGURE 29.2 A repeat EEG recorded during sleep, demonstrating normal sleep architecture and lack of spike–wave activity.

parents noted improvement, but he did not return to his previous cognitive baseline. The continuous spike–wave activity during sleep improved (see Figure 29.2). A 2-Deoxy-2-[18F] fluoro-D-glucose (FDG) positron emission tomography (FDG-PET) scan was interpreted as normal. Prednisone was tapered and discontinued on two occasions with worsening of his school performance. Lamotrigine was substituted for VPA without any clinical improvement. Currently, he exhibits only rare seizures on prednisone and lamotrigine therapy.

DIFFERENTIAL DIAGNOSIS

In the differential diagnosis, two aspects of CSWS should be considered: the typical EEG findings during sleep, and the acute onset of cognitive and behavioral regression. With respect to EEG changes, patients with Lennox–Gastaut syndrome (LGS) may also show nearly continuous slow spike-and-wave discharges during sleep (see Chapter 25). However, patients with LGS are often developmentally delayed from early childhood, and usually do not have an abrupt regression as seen in CSWS. Further, LGS patients often exhibit a combination of myoclonic, tonic, atonic, and atypical absence seizures that are difficult to control. Another syndrome characterized by abundant epileptiform discharges during sleep is benign childhood epilepsy with centrotemporal spikes (aka, benign Rolandic epilepsy). Clinically, these patients may have learning disability or attention deficit hyperactivity disorder (ADHD), but significant regression is not seen. Finally, in the differential of CSWS is Landau–Kleffner syndrome (LKS; see Chapter 28), which is characterized by focal, not generalized, nearly continuous sleep-related spike–wave discharges, seen primarily in the dominant temporal lobe. Furthermore, patients with LKS have significant language regression, but sparing of other cognitive abilities.

Epilepsy syndromes that are associated with regression include the progressive myoclonic epilepsies, such as Lafora body disease or Unverricht–Lundborg disease. In these cases, the regression is often associated with worsening generalized myoclonic seizures, which were not present in the representative case described. Neuronal ceroid lipofuscinoses (NCLs) may also present with similar symptoms, but this set of progressive myoclonic epilepsies is often associated with visual impairment. It is important to note that cognitive slowing may be a side effect of antiepileptic medications, particularly when administered at high doses, irrespective of the epilepsy syndrome.

DIAGNOSTIC APPROACH

The clinical entity of CSWS, also called electrical status epilepticus of sleep (ESES), was initially described by Patry et al. in 1971. CSWS is diagnosed on the basis of the percentage of the sleep EEG recording comprising spike–wave activity. The traditional definition requires that the spike–wave index be greater than 85% during slow-wave sleep. In addition to the EEG criteria, the appearance of neuropsychological regression is required during the period of CSWS activity.

3. Treatment with antiepileptic medications often has little impact on the EEG findings in CSWS, whereas treatment with steroids and high-dose benzodiazepines has been shown to be effective.

4. The long-term neuropsychological prognosis for CSWS is guarded, with nearly 50% of patients experiencing continued difficulties despite resolution of the EEG pattern during adolescence.

SUGGESTED READING

Galanopoulo AS, Bojko A, Lado F, Moshé SL. (2000) The spectrum of neuropsychiatric abnormalities associated with electrical status epilepticus in sleep. *Brain Dev.* 22: 279–295.

Guzzetta F, Battaglia D, Veredice C et al. (2005) Early thalamic injury associated with epilepsy and continuous spike–wave during slow sleep. *Epilepsia* 46(6): 889–900.

Inutsuka M, Kobayashi K, Oka M, Hattori J, Ohtsuka Y. (2006) Treatment of epilepsy with electrical status epilepticus during slow sleep and its related disorders. *Brain Dev.* 28: 281–286.

Nieuwenhuis L, Nicolat J. (2006) The pathophysiological mechanisms of cognitive and behavioral disturbances in children with Landau–Kleffner syndrome or epilepsy with continuous spike-and-waves during slow-wave sleep. *Seizure* 15: 249–258.

Smith M, Polkey C. (2008) Landau–Kleffner syndrome and CSWS. In *Epilepsy: A Comprehensive Textbook, 2nd ed.*, Engel J, Pedley T, Eds., 2429–2437. Philadelphia: Lippincott Williams & Wilkins.

Tassinari CA, Rubboli G, Volpi L et al. (2000) Encephalopathy with electrical status epilepticus during slow sleep or ESES syndrome including the acquired aphasia. *Clin. Neurophysiol.* 111: S94–S102.

30 Rasmussen Encephalitis

*Daniel H. Arndt, M.D. and Raman
Sankar, M.D., Ph.D.*

CONTENTS

CASE PRESENTATION

A right-handed boy experienced his first seizure at 4 years of age while playing with his mother at home. The mother noted leftward pulling of his face, followed by an arrest of behavior, staring, chewing automatisms, drooling, and unresponsiveness for 5 minutes. No postictal weakness was observed despite some transient lethargy. He was afebrile and without any symptoms or signs of infection. Birth history and early development were normal. The family denied any prior history of febrile seizures, meningitis/encephalitis, traumatic brain injury, unexplained loss of consciousness, or family history of epilepsy. His neurological examination was normal for age. Interictal electroencephalogram (EEG) showed infrequent R frontoparietal epileptiform discharges, and his magnetic resonance imaging (MRI) was normal. He was placed on oxcarbazepine therapy, but his complex partial seizures became more frequent and severe. Consequently, two other antiepileptic medications were tried over the next 3 months without added benefit. Semiology had progressed to include secondary generalization, and convulsions became frequent. All seizures were now preceded by an aura of left perioral or hemibody pain. Valproic acid (VPA) was effective in controlling complex partial seizures, but he continued to experience 1–2 typical auras per week. Additionally, he displayed a mild, progressive, left hemiparesis and some neurocognitive decline. Nearly 4 months after seizure onset, a repeat MRI showed new hypointense T1 and hyperintense T2 signals along the R perisylvian gray and white matter. Spinal fluid examination was unremarkable. Prolonged video-EEG monitoring was performed, and interictal EEG tracings showed frequent, multifocal right hemispheric epileptiform discharges centered over the right central and temporal areas (T4, T6, C4, and P4). Subclinical and clinical seizures were observed to evolve from the F4 lead, some with secondary generalization. The clinical component included left hand and foot *epilepsia partialis continua*, along with left face and hemibody complex partial seizures of 1–2 minute duration. Interictal Fluoro-D-glucose positron emission tomography (FDG-PET) showed right frontal, temporal, deep insular, basal ganglia, and thalamus hypermetabolism, as well as left inferior cerebellar hypermetabolism. Some right temporal hypometabolism was also present. He was given the presumptive diagnosis of Rasmussen encephalitis (RE). After discussion of the various therapeutic options, a right hemispherectomy was performed. Gross inspection of the cortex and white matter did not reveal recognizable malformations. However, microscopic sections confirmed the suspected clinical diagnosis of Rasmussen encephalitis, revealing loss of neurons with marked patchy astrogliosis, microglial nodules scattered throughout the cortex, and perivascular cuffing with lymphoid cell infiltrates. He has been seizure free for 3½ years since surgery, though maintained on a modest dose of levetiracetam (25 mg/kg/day). He continues with a mild–moderate hemiparesis, and he is ambulatory with a hemiparetic gait. Neurocognitive function and language are age appropriate.

DIFFERENTIAL DIAGNOSIS

In the early stages, RE can begin with partial seizures with focal epileptiform activity on the EEG, whereas the MRI may be normal. The initially normal MRI differentiates this syndrome from other symptomatic partial epilepsies such as focal cortical dysplasia, cysticercosis, malignancies, or cortical injury-related etiologies. RE lacks the specific association of sleep-evoked clinical and EEG features, as well as the surface dipole seen in idiopathic partial epilepsy of childhood (Rolandic) epilepsy. The early presentation of RE consists of partial seizures without the encephalopathy that one would associate with acute encephalitis.

   RE is a syndrome of intractable partial epilepsy, progressive hemiparesis, and characteristic histopathologic changes in the brain. Although the actual incidence is unclear, it is rarely encountered in common clinical practice. The disease begins before 10 years of age in 85% of cases, though cases may begin during later adult life. The initial phase of the disease may involve rare, easy-to-control partial seizures that can last for a few months to years. The second phase is associated with an acceleration of the disease, with recurrent partial seizures and unilateral neurologic dysfunction. This most commonly involves a progressive hemiparesis, but other dysfunction, such as language and cognitive dysfunction, can occur, depending on whether the affected hemisphere is the dominant one. This phase is often rapid and may last a few months to a year, and often is associated with hospitalization and referral to an epileptologist. The third phase often describes a point at which the neurologic dysfunction plateaus and the seizure frequency typically lessens.

DIAGNOSTIC APPROACH

The diagnosis of RE involves a congruence of clinical history with typical neuro-physiologic and neuroradiologic changes. The classic neuropathologic features may be demonstrated by a brain biopsy and are confirmatory, but in most patients this is not considered necessary. The EEG abnormalities are progressive, and early studies can even be normal. The affected hemisphere characteristically shows slowing of the background, disrupted sleep architecture, and frequent, sometimes continuous, epileptiform discharges. *Epilepsia partialis continua* involves continuous or almost continuous focal, rhythmic clonic activity that may persist during sleep and can last from hours to weeks and occurs in 50% of RE patients.

The predominant finding on MRI is perisylvian cortical atrophy of the involved hemisphere, and may overlap with T2/FLAIR hyperintensity (Figure 30.1 A and B). Serial imaging during the acute phase shows progressive changes unilaterally, and possibly bilateral in rare and advanced cases. FDG-PET imaging typically identifies hypermetabolic areas superimposed on hypometabolic regions. Focal or diffuse hypometabolism may precede MRI abnormalities (Figure 30.1 C and D).

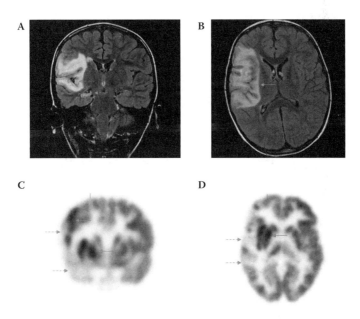

FIGURE 30.1   (A) Coronal FLAIR MRI showing classic T2/FLAIR perisylvian hyperintensity in gray and white matter, with sylvian fissure enlargement due to gyral atrophy (solid arrow). Mesiotemporal sclerosis is also present (dashed arrow). (B) Corresponding Axial FLAIR MRI. (C) Coronal and (D) Axial FDG-PET showing mixed R temperoparietal hypometabolism (dashed arrow) and hypermetabolism (solid arrow), and R basal ganglia hypermetabolism (solid arrow).

Neuropathology classically shows neuronal loss, perivascular lymphocytic cuffing, and proliferation of microglial nodules in the affected cortex. These findings, taken with the clinical picture, are confirmatory of RE. Importantly, these findings rule out other considerations in the differential diagnosis such as focal cortical dysplasia. With typical EEG, neuroimaging, and clinical history, brain biopsy is usually unnecessary.

TREATMENT STRATEGY

Initial treatment involves antiepileptic drugs (AEDs) known to be effective in typical localization-related epilepsies. However, most cases rapidly become medically refractory, and seizures worsen despite progressive escalation in dose and number of AEDs. Treatments advocated in the literature include AEDs, high-dose steroids, ACTH, intravenous immunoglobulin, plasmapharesis, and hemispherectomy. Although immunologic modulation with steroids or intravenous immunoglobulin may transiently improve symptoms or slow the progression of disease, it is rarely curative. Hemispherectomy is indicated for unilateral RE, and can be curative. Hemispherectomy should be discussed with the family early in the course of RE, and is often advocated once hemiparesis is apparent or worsening seizures significantly impact the patient's quality of life.

LONG-TERM OUTCOME

The natural history of RE suggests a progressive decline in neurologic function associated with worsening epilepsy and unilateral hemispheric atrophy to a plateau point with stabilization of function and fewer seizures. Hemispherectomy offers the best chance of seizure freedom in unilateral RE cases. Studies have shown that 88% of hemispherectomy patients became seizure free or had occasional, nondisabling seizures. There are very rare cases of unilateral RE developing into bilateral disease, and hemispherectomy *may* prevent progression of the disease to the contralateral hemisphere. Early hemispherectomy is advocated despite an expected worsening of hemiparesis. With early restoration of seizure control, the child has the opportunity to regain developmental losses and achieve an improved neurocognitive outcome. Hydrocephalus occurs as a complication of surgery in about one-fourth of patients.

PATHOPHYSIOLOGY/NEUROBIOLOGY OF DISEASE

The etiology of RE is unclear, but an autoimmune process is suspected. A relationship to antibodies to the AMPA receptor GluR3 subunit was proposed on the basis of the incidental discovery of a rabbit model of severe partial seizures and an encephalitis-like picture. However, not all RE cases have shown GluR3 autoantibodies, and these same antibodies have been seen in resected tissue in other noninflammatory epilepsies. The neuropathologic changes of perivascular lymphocytic cuffing and proliferation of microglial nodules in the affected cortex suggest an immune or autoimmune basis. Postinfectious and parainfectious etiologies have been speculated, based on sporadic detection of CMV and HSV1 by PCR and *in situ* hybridization; however, these findings have not been consistently duplicated. Therefore, an autoimmune process is still suspected, but the specific provocation or the mechanism that selects hemispheric involvement of the disease is still unclear.

CLINICAL PEARLS

1. Rasmussen encephalitis is a focal, progressive, inflammatory brain condition with unilateral onset.
2. Clinically, children develop intractable focal motor seizures and secondarily generalized seizures, progressive hemiparesis, progressive visual field deficits, and declining cognitive performance.
3. MRI, EEG, and PET show progressive unilateral hemispheric progression.
4. Neuropathology shows neuronal loss, perivascular lymphocytic cuffing, and proliferation of microglial nodules in the affected cortex.
5. Initial treatment involves typical AEDs directed against localization-related epilepsy, but once a clinical diagnosis of unilateral Rasmussen encephalitis is strongly suspected, a discussion of early hemispherectomy is appropriate, and such surgery can be curative.

SUGGESTED READING

Andermann F, Ed. (1991) *Chronic Encephalitis and Epilepsy: Rasmussen Syndrome.* Boston: Butterworth-Heinemann.

Bien C, Widman G, Urbach H. et al. (2002) The natural history of Rasmussen's encephalitis. *Brain* 125: 1751–1759.

Freeman J. (2005) Rasmussen's syndrome: progressive autoimmune multi-focal encephalopathy. *Pediatr. Neurology* 32: 295–299.

Hart Y, Cortez M., Andermann F. et al. (1994) Medical treatment of Rasmussen's Syndrome (chronic encephalitis and epilepsy): effect of high-dose steroids or immunoglobulins in 19 patients. *Neurol.* 44(6), 1030–6.

Kotagal P. (2008) Localization-related epilepsy: simple partial seizures, complex partial seizures, and Rasmussen syndrome. *Pediatr. Epilepsy: Diagnosis and Treatment,* 3rd ed., Pellock, J. M., Dodson, W. E., Bourgeois, B. F. D., Nordli, Jr., D. R., and Sankar, R. (Eds.), 377–385. New York: Demos Medical Publishing.

Vining E, Freeman J, Pillas D. et al. (1997) Why would you remove half a brain? The outcome of 58 children after hemispherectomy—the Johns Hopkins experience: 1968 to 1996. *Pediatrics* 100 (2): 163–171.

31 Myoclonic–Astatic Epilepsy

A.G. Christina Bergqvist, M.D.

CONTENTS

CASE PRESENTATION

FR is a 3-year-old boy with a normal birth, development, and past medical history. At age 3 he had his first generalized tonic–clonic seizure during a viral illness accompanied by a high fever. He was given a diagnosis of febrile seizures. In the next months, he had five convulsions without fever and was evaluated by a neurologist. He was given a diagnosis of partial seizures and was started on carbamazepine, which exacerbated the convulsions. A workup including an MRI of his brain and a metabolic screen (serum aminoacids, urine organic acids, lactate and pyruvate, and acylcarnitine esters) was normal. An initial electroencephalogram (EEG) showed biparietal theta but no epileptiform discharges. He was switched to valproic acid (VPA).

Over the next year, he developed multiple seizure types, myoclonic–astatic seizures with myoclonus of the shoulder and head drops. He had two bouts of nonconvulsive status epilepticus (SE) and absences that could be prolonged. The EEG continued to show biparietal slowing, but now bifrontal 2–3 Hz spike–wave discharges were also present. One video EEG monitoring session capturing the child's convulsion showed generalized discharges at the seizures onset. An ophthalmology evaluation of his retina was normal.

Medical treatment trials continued: VPA was pushed to 125 mg/dL as a monotherapy, and chlorazepate was added, which reduced the convulsive seizures but resulted in hyperactivity and drooling; lamotrigine was then tried and pushed to a level of 12 mg/dL without change in his seizure control. Zonegran improved the myoclonic seizures briefly, but they later returned. Convulsive seizures continued daily; absences and myoclonic–astatic seizures were seen throughout the day. Up to this point he had been a normally developing child, but now his language regressed to where he was speaking in one- or two-word sentences or pointing. He was drooling. He would spend most of his day sitting, surrounded by pillows to avoid falls and head injury. His motor function was also affected, and he walked wide-based and appeared ataxic. At age 4.5 years, he was started on the ketogenic diet (KD). Within 1 week of KD therapy, all of his seizures abated. Within 6 months of KD therapy, he was successfully weaned off all his medications, and his language and balance returned to normal. After 3 years of KD therapy, he was successfully weaned onto regular food without recurrence of his seizures. He is currently 8 years old and attends third grade in regular classes.

DIFFERENTIAL DIAGNOSIS

This child has a diagnosis of myoclonic astatic epilepsy (MAE), sometimes referred to as Doose syndrome after the initial describing author. MAE is classified as a generalized epilepsy, which develops in normal children between the ages 2 to 3 years (rarely as early as at 1 year of age). MAE is more common in boys than in girls by about a 2:1 to 3:1 ratio. It is characterized by multiple seizure types, predominantly myoclonic, astatic (drop attacks), myoclonic–astatic seizures, generalized tonic–clonic seizures, absences, and myoclonic absences seizures. Myoclonic–astatic seizures are the characteristic seizure type in this syndrome and typically consist of a symmetric myoclonic jerk followed by atonia causing a fall if standing. Generalized tonic–clonic seizures usually herald the syndrome, and every child with MAE will develop myoclonic–astatic seizures. Rarely, nocturnal generalized tonic seizures may develop. The differential diagnosis includes severe myoclonic epilepsy of infancy (SMEI; also known as Dravet syndrome), myoclonic Lennox–Gastaut syndrome (LGS), and progressive myoclonic epilepsies. The differentiation between these disorders can be difficult and is based on developmental status prior to seizures, age of onset, seizures types and EEG patterns, as well as the absence of metabolic findings and a normal magnetic resonance imaging (MRI).

Myoclonic LGS is the most significant differential diagnosis. There are several differences that should help distinguish between the two syndromes. Children with LGS do not have a normal developmental history before the onset of the seizures. Tonic seizures predominate, and they are partial with secondary generalization when viewed on EEG. Atypical absences are prominent with LGS, which are not seen with MAE. Children with LGS also may have partial discharges and activation of generalized discharges during slow-wave sleep, which is not a characteristic of MAE. LGS is classically associated with slow spike–wave activity on EEG, although this activity may occasionally be seen in MAE.

SMEI is also a generalized epilepsy, but can be differentiated from MAE by earlier onset of seizures (in the first year of life), often associated with fever. Although myoclonic seizures predominate, they may also have generalized tonic–clonic convulsions and hemiconvulsions. Children with SMEI often have a defect in the sodium channel gene (*SCN1A*). The cognitive outcome is almost universally affected negatively, although approximately half of the children with MAE have "normal" cognitive function once their seizures abate.

Occasionally, progressive encephalopathies (e.g., mitochondrial disorders, late-infantile neuronal ceroid lipofuscinosis) may mimic the clinical course, but in general, myoclonic seizures predominate. These disorders may also be distinguished through biochemical or other testing that identifies the underlying etiology.

DIAGNOSTIC APPROACH

A good history focusing on the development and the seizure types as they emerge is essential to making the correct diagnosis. Video-EEG monitoring can be very helpful in documenting the seizure types. Initial EEGs are often normal or may show a characteristic bicentral or parietal theta before the epileptiform discharges emerge. Generalized 2–3 Hz or irregular polyspike discharges then evolve. Myoclonic seizures are associated with a burst of 2–4 Hz spike–wave or polyspike activity and involve symmetric myoclonus, most often involving the proximal upper extremities. Partial discharges on initial EEG are an exclusion criteria, as should be onset of seizure before age 1 and daytime tonic seizures.

Children with MAE often become encephalopathic when their seizures are frequent. Progressive encephalopathies and myoclonic seizures due to inborn errors of metabolism should be ruled out by metabolic screenings (serum lactate and pyruvate, plasma amino acids, urine organic acids, and acylcarnitine esters). MRI/magnetic resonance spectroscopy (MRS) imaging is normal in MAE and remains so during their disease course.

TREATMENT STRATEGIES

The optimal treatment of MAE is not known. There has been no randomized clinical trial comparing our current treatment options. Treatment is therefore extrapolated from our knowledge of treating idiopathic generalized epilepsies.

Many children with MAE are initially erroneously treated with oxcarbazepine or carbamazepine (initial diagnosis thought to be partial epilepsy), which will worsen their seizures. Antiepileptic drugs that have shown promise in small clinical series include the following: VPA, ethosuximide, topiramate, lamotrigine, and levetiracetam. These antiepileptic drugs (AEDs) are often combined with benzodiazepines, though side effects and tolerance may limit their utility. Others that have been used with varied success include zonisamide, steroids, felbamate, vigabatrin, and intravenous immunoglobulin.

KD, a 90% fat, 7% protein and 3% carbohydrate diet that mimics the metabolic changes that occur with fasting, has shown promising efficacy in the MAE population, offering 30–50% seizure freedom. The KD is often used as a last resort and should possibly be tried sooner in these children. Although studies have demonstrated the beneficial effect of vagus nerve stimulation in atonic seizures, no studies have directly addressed its utility in MAE. Corpus callosotomy and other epilepsy surgeries are not indicated.

LONG-TERM OUTCOME

Long-term prognosis in the MAE group is variable, and ranges from seizure freedom with normal development to severe retardation and continued refractory epilepsy. Doose, in the largest series of MAE children, reported that about half of the children older than 7 became seizure free for at least 2 years or more. Others have found slightly higher remission rate in the range of 60–70%. Seizure recurrence years after initial remission has been reported, but is rare. These seizures, should they occur, are usually easily controlled. A relationship to poor control of seizures and cognitive deterioration has been hypothesized, but not proved.

PATHOPHYSIOLOGY/NEUROBIOLOGY OF DISEASE

The etiology for MAE is unknown, but a genetic mechanism is suspected. The male predominance and high incidence of a family history of epilepsy or febrile seizures support the genetic hypothesis. Many series report 30–50% family history of febrile seizures or epilepsy. Relationships to the *GEFS+* genes have been suggested, but the data presented are inconclusive. It is likely that the MAE phenotype results from several genetic abnormalities, and modifier genes and environmental factors may play additional roles.

CLINICAL PEARLS

1. MAE is a diagnosis of exclusion, and should be supported by the proper documentation of the initially normal developmental history, seizure type and evolution, EEG findings, absence of structural changes on MRI, and a normal metabolic screen.
2. Treatment should be aggressive and initially include use of AEDs used for generalized epilepsy.
3. The ketogenic diet should be considered earlier in the treatment of these children and offered to the family once the MAE diagnosis is made.
4. Prognosis for MAE is variable with about one half of patients achieving seizure freedom and near-normal IQ.

SUGGESTED READING

Kilaru S, Bergqvist AG. (2007). Current treatment of myoclonic astatic epilepsy: clinical experience at the Children's Hospital of Philadelphia. *Epilepsia* 48(9), 1–5.

Doose H. (1992). Myoclonic-astatic epilepsy. *Epilepsy Res.—Supplement.* 6, 163–168.

Guerrini R, Aicardi J. (2003). Epileptic encephalopathy with myoclonic seizures in infants and children (severe myoclonic epilepsy and myoclonic-astatic epilepsy). *J. Clin. Neurophysiol.* 20, 449–461.

Neubaur BA, Hahn A, Doose H, Tuxhorn I. (2005). Myoclonic-astatic epilepsy of early childhood-definition, course, nosography and genetics. *Adv. Neurol.* 95, 147–155.

Oguni H, Tanaka T, Hayashi K. et al. (2002). Treatment and long-term prognosis of myoclonic-astatic epilepsy of early childhood. *Neuropediatrics* 33, 122–132.

Section 5

The Adolescent

32 Juvenile Myoclonic Epilepsy

Cornelia Drees, M.D.

CONTENTS

CASE PRESENTATION

The patient is a 15-year-old boy who was seen at an emergency department in the early morning hours after he was witnessed to have a seizure at a party. He was at summer camp and had stayed up longer—and slept less—than usual for most of the preceding days. His friends recalled that he yelled, stiffened, and then jerked with his whole body for about 1 minute. He had bloody frothing at the mouth and lost bladder control. Afterwards, the staff at the camp was unable to arouse him for another 15 minutes, and he later woke up slightly disoriented, complaining of sore muscles "all over." The patient did not recall any warning signs prior the incident, and was amnestic for the episode. This was the first event of this kind, and he and his parents denied any prior history of seizures or staring spells during infancy and childhood. However, when asked specifically, he admitted to noticing muscle twitches in his shoulders and arms for the past 6 months. They typically occurred in the morning, after getting up, and had, at times, interfered with his daily routine before leaving for school. He recalls once involuntarily throwing a toothbrush across the bathroom, and another time when a jerk made him drop his cereal bowl on the floor. His examination was completely normal.

DIFFERENTIAL DIAGNOSIS

A seizure in an adolescent should always raise concern for a provoked seizure, that is, elicited by a trigger, such as sleep deprivation, alcohol withdrawal after alcohol excess, withdrawal from benzodiazepines, or use of illicit drugs such as amphetamines or cocaine.

Primary generalized epilepsies that first manifest with seizures during the teenage years should also be considered, such as juvenile myoclonic epilepsy (JME) or juvenile absence epilepsy (JAE). Childhood absence epilepsy may present with a generalized seizure, but it is unusual for associated staring spells to go undetected by parents and teachers for years, and typically seizures remit during puberty. Obviously, a seizure could have occurred owing to an underlying structural lesion, such as a cortical dysplasia, a vascular malformation, or a tumor, something to remember even when acute provocation seems most likely (e.g., from sleep deprivation or intoxication).

DIAGNOSTIC APPROACH

As always, the key to the diagnosis lies in the history! This young man experienced vigorous myoclonic jerks while completely awake. Myoclonic seizures have a propensity to occur in the early morning hours. In combination with a generalized tonic–clonic seizure, the most likely diagnosis is JME. Typical EEG findings confirm this suspicion, that is, 4–6 Hz spike–wave and polyspike–wave complexes, occurring spontaneously out of an otherwise normal background (see Figure 32.1). In approximately 30% of patients, these discharges are triggered by photic stimulation, and some will appear after provocation by sleep deprivation or drinking caffeinated beverages. During a myoclonic seizure, the myoclonic jerk corresponds to a polyspike–wave discharge. In contrast, other myoclonic jerks, for example, when an individual is drifting off to sleep (hypnagogic myoclonic jerk or sleep myoclonus) or when startled, are not accompanied by epileptiform activity. Neuroimaging in a typical case is not required, because the brain is structurally normal in these patients. However, many physicians will obtain a magnetic resonance imaging (MRI) of the brain to rule out other reasons for seizures that may mirror the presentation. It should be noted that the EEG of JME patients may exhibit focal features, thus suggesting a structural anormality or lesion in the brain. A toxicology/alcohol screen should be done, if there is a suspicion that drugs might be involved.

TREATMENT STRATEGY

Patients with JME have greater than 90% chance of experiencing recurrent seizures. Therefore, lifelong treatment with antiepileptic drugs (AEDs) and avoidance of possibly provoking factors (e.g., alcohol, illicit drugs, sleep deprivation, and flickering lights) are recommended. Therapeutic agents effective against seizures in generalized epilepsies should be used and are usually very effective in controlling seizures. Currently, lamotrigine, levetiracetam, topiramate, zonisamide, and valproic acid (VPA in males), are considered first-line treatment. VPA is typically avoided

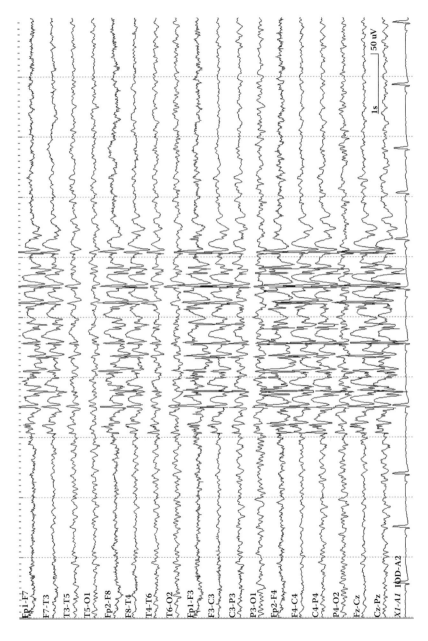

FIGURE 32.1   EEG demonstrating 4–6 Hz generalized spike–wave and polyspike–wave discharges.

in young women because of cosmetic and hormonal adverse effects, teratogenicity, and an increased likelihood of cognitive problems in children who were exposed to it *in utero*. Long-acting benzodiazepines, such as clonazepam or clobazam, are also effective, but have the potential for tolerance and addiction. Most patients (80–90%) are controlled on a single agent (i.e., monotherapy), but some require combination therapy. Several drugs used for partial-onset seizures can exacerbate seizures and even cause status epilepticus in patients with JME (e.g., phenytoin, carbamazepine, oxcarbazepine, gabapentin, tiagabine, and vigabatrin), and should not be used.

LONG-TERM OUTCOME

Studies on the natural history of the disorder indicate that virtually all patients will have recurrent seizures after antiepileptic medication is discontinued. Unfortunately, this mandates that JME patients need to be treated for life. However, despite this poor prognostic fact, they have otherwise normal intelligence and life expectancy. Generally speaking, it is estimated that a mother with idiopathic generalized epilepsy has about a 10% risk of having a child with generalized epilepsy.

PATHOPHYSIOLOGY/NEUROBIOLOGY OF DISEASE

JME is strongly genetically linked and several genes have been implicated that, when mutated, can alone or in combination cause the disorder. Although the family history is positive in up to 50% of patients, the inheritance patterns are often not clear-cut and suggest a process influenced by multiple genes and possibly other factors. The net result of one or multiple gene defects in combination with other factors leads to increased neuronal excitability, especially a tendency for thalamocortical networks to produce spike–wave and polyspike–wave complexes seen on EEG.

Gene mutations on chromosomes 2, 3, 5, 6, 15, and 18 are associated with JME, and in the future this list may grow. However, the family with a monogenic form of JME is a rarity. Identified pedigrees have obviously been studied extensively and are those that tell us something more specific about the connection between specific gene defects and the electroclinical manifestation of epilepsy.

For example, it was recently reported that some of these gene defects give rise to so-called "channelopathies"—conditions known to alter normal ion channel function which can yield an episodic clinical phenotype such as epilepsy. A mutation on chromosome 3q, for instance, affects the voltage-gated chloride channel (i.e., *ClCN2*), and another mutation on this chromosome is associated with changes in the calcium-activated potassium channels. In Indian families with JME, a gene on chromosome 8q coding for a potassium channel has been implicated. As another example, an autosomal dominant mutation of a gene on chromosome 5q (i.e., *GABRA1*) interferes with the production of an important $GABA_A$ receptor subunit, which in turn alters the effectiveness of the inhibitory transmitter Gamma-aminobutyric acid (GABA) on postsynaptic neurons. An altogether different mechanism is suspected for a defect in a gene called *EFHC1*, which changes the function of certain voltage-dependent calcium channels normally involved in neuronal apoptosis during early brain development. It is speculated that by *not* eliminating some cells, a hyperexcitable circuit is fostered.

CLINICAL PEARLS

1. JME patients present with the following seizure types, in isolation or in combination: (a) absence seizures (i.e., staring spells), (b) myoclonic seizures (brief muscle jerks), and (c) generalized tonic–clonic seizures.
2. Myoclonic seizures occur early in the disease course and very typically occur in the morning, although this feature is often overlooked until the patient has a generalized tonic–clonic seizure.
3. Myoclonic seizures—also called "jerks," "twitches," "electricity," "shocks," or "like lightning"—can occur in prolonged clusters without significant impairment of consciousness, or they can build up to a generalized tonic–clonic seizure.
4. A positive family history is not uncommon and is found in 10–50% of patients, although first-degree relatives are not necessarily more likely to have seizures than other family members. A few relatives have only characteristic EEG findings, but no seizures.
5. JME patients should be warned that they will most likely be on lifelong treatment with antiepileptic drugs. However, all first-line antiepileptic drugs readily offer seizure control and seizure freedom in 80–90% of patients, and thus, life is fairly normal for most patients.

SUGGESTED READING

Andermann F, Berkovic S. (2004). The idiopathic generalized epilepsies across life. *Clin. Neurophysiol.* 57 (Suppl.) 408–414.

Dinner DS, Lüders H, Morris HH, Lesser RP. (1987). Juvenile myoclonic epilepsy. In *Epilepsy: Electroclinical Syndromes*, K. Levin, H. Lüders, R. P. Lesser (Eds.), 131–149. London: Springer Verlag.

Panayiotopoulos CP. (2005). Idiopathic generalized epilepsies: A review and modern approach. *Epilepsia*, 46 (Suppl.) 9, 10–168.

Glauser T, Ben-Menachem E, Bourgeois B. et al. (2006). ILAE treatment guidelines: evidence-based analysis of antiepileptic drug efficacy and effectiveness as initial monotherapy for epileptic seizures and syndromes. *Epilepsia* 47, 1094–1120.

Martinez-Juarez IE, Alonso ME, Medina MT. et al. (2006). Juvenile myoclonic epilepsy subsyndromes: family studies and long-term follow-up. *Brain* 129 (pt. 5), 1269–1280.

Zifkin B, Andermann E, Andermann F. (2005). Mechanisms, genetics, and pathogenesis of juvenile myoclonic epilepsy. *Curr. Opin. Neurol.* 18, 147–153.

33 Epilepsy in Adolescent Females

Mary L. Zupanc, M.D.

CONTENTS

CASE PRESENTATION

The patient is a previously healthy 14-year-old adolescent with a new-onset generalized tonic–clonic seizure. She was sitting at her computer in the early evening hours after having been to a slumber party with her friends the evening before. Suddenly, she felt somewhat "dizzy" and confused. This was followed quickly by a generalized tonic–clonic seizure without focal features. The seizure lasted approximately 5 minutes and was followed by a 20-minute period of postictal lethargy and confusion. A witness called 911, and she was taken by ambulance to the local emergency room. Her physical examination was normal. Routine blood chemistries were performed, along with a toxicology screen. A head computed tomography (CT) scan was also performed. All studies were normal. She was advised to be seen in follow-up by a neurologist.

DIFFERENTIAL DIAGNOSIS

This patient had a single unprovoked generalized tonic–clonic seizure, in the context of sitting in front of a computer, with possible drowsiness. The computer raises the possibility of photic sensitivity, which can be seen with generalized epilepsy syndromes. However, the patient did complain of dizziness prior to the onset of her generalized tonic–clonic seizure. This could represent a simple partial seizure (i.e., aura) followed by secondary generalization. Alternatively, this could imply that the patient had a brief series of generalized 3–4 Hz spike-and-slow-wave discharges, which in

245

discourage baths and encourage shower, due to increased risk of drowning. Driving is permitted in all states after a variable period of complete seizure freedom—typically the waiting period is 6 months to 1 year seizure-free on AEDs.

Compliance with antiepileptic medication is a challenge in all patients with epilepsy, but more so during adolescence. Studies suggest that the biggest risk factor for noncompliance is the age of adolescence. Although parents typically administer medication to younger children, at some point this process undergoes a transition, and the responsibility of compliance begins to shift to the teenager. In studies of adolescents with chronic disorders, compliance is reported to be incomplete in 50% of patients. Longer duration of illness, less exercise, smoking and alcohol use, and more frequent seizures were associated with poorer compliance. Strategies to improve compliance include recognizing and exploring the problem with the patient, suggesting AEDs with twice-daily or once-daily dosing, and helping the adolescent set up a routine (bedtime, mealtimes, etc.) for taking the medication.

Lack of adequate sleep is a growing problem among American teenagers, as teens juggle rigorous academic demands, busy social lives, part-time jobs, and late nights at the computer. Nowhere is this more relevant than for the adolescent with epilepsy, for whom sleep deprivation may provoke a seizure, as with our patient. The association between sleep deprivation and epilepsy is well documented, particularly in the setting of the idiopathic generalized epilepsies.

With the teenage patient as the focus of consultation rather than the parents, the patient is more likely to reveal and discuss concerns. Although a specialized adolescent epilepsy clinic may not be feasible in many centers, the strategies employed are applicable to any clinical setting. The practitioner is advised to consider a comprehensive checklist for a visit involving an adolescent with epilepsy (see Table 33.1).

LONG-TERM OUTCOME

Any patient who has the onset of epilepsy during adolescence should be counseled that the epilepsy is probably a chronic condition, one that is unlikely to go into remission. For our patient, JME is known as a lifelong condition. This is often very difficult for adolescents to comprehend. The diagnosis of epilepsy can result in

---

TABLE 33.1
Checklist for adolescent epilepsy visit

Seizure control

Medication side effects: emphasis on cognitive slowing and drowsiness, and weight

Nutrition

School performance and attendance, participation in sports

Signs of depression

Sleep deprivation

Contraception

Recreational drugs and alcohol

anxiety and depression. The risk of suicide in adolescents with epilepsy is higher than that in the general population.

NEUROBIOLOGY

JME and other epilepsy syndromes present at the onset of adolescence, in the midst of menarche and the other changes associated with puberty. Although most forms of idiopathic, generalized epilepsy syndromes are believed to arise from as-yet poorly defined seizure susceptibility genes, hormonal changes during the menstrual cycle can exert a profound effect on seizure activity. Estrogen is a potent proconvulsant, whereas progesterone has anticonvulsant effects. The ratio of these two hormones probably influences the tendency of breakthrough seizures. With some young adolescent women, the highest risk for breakthrough seizures is either at the time of ovulation or right before menses, when the estrogen:progesterone ratio is at its peak. For other adolescent women, in particular those with anovulatory cycles, the risk for breakthrough seizures may persist throughout the cycle because of the unopposed action of estrogen.

Catamenial seizures are a real problem for many young adolescent women with epilepsy. Estimates of women with catamenial seizures vary greatly in the literature, but the incidence has been reported to be as high as 75% of all women with epilepsy, including young adolescent women. Treatment for catamenial seizures has not been well studied. However, preliminary studies indicate that progesterone lozenges or suppositories may be helpful in the prevention of catamenial seizures. Other recommended treatments for catamenial seizures have included acetazolamide (Diamox) and oral contraceptives.

CLINICAL PEARLS

1. Adolescent women can present with epilepsy at the onset of puberty, and it can become a chronic, lifelong condition. One of the most common epilepsy syndromes is JME.
2. Adolescence is a time of great change—both physically and emotionally. If an adolescent develops epilepsy, the challenges are considerable, and there are higher risks for comorbid anxiety, depression, and suicide.
3. Adolescent young women with epilepsy are at risk for reproductive dysfunction, including anovulatory cycles, menstrual irregularities, polycystic ovarian syndrome, and sexual dysfunction. Antiepileptic medications can actually increase these risks, particularly VPA. As such, VPA is relatively contraindicated in women with epilepsy who are at a reproductive age.
4. AEDs do carry teratogenic risks. These risks should be fully disclosed to any adolescent young woman and her family during the discussion about AED choices. Also, there is the additional risk of breakthrough seizures during pregnancy.

5. AEDs do have hormonal interactions. In particular, those that induce the P450 enzyme system metabolize the steroid hormones more rapidly. Therefore, if oral contraceptives are used in combination with AEDs, the dose of estrogen in the OCP should be at least 50 micrograms. If depo-provera is used, the injections should be given more frequently. VPA inhibits the P450 enzyme system.

6. Nutrition is generally less than optimal in adolescents, so it is vital to emphasize the importance of adequate calcium intake, folate supplementation, and overall good nutritional habits. Young adolescent women with epilepsy may be at greater risk for osteoporosis than their peers.

7. Sports are not contraindicated in individuals with epilepsy.

8. An adolescent clinic for young women with epilepsy is the best way to address many of these issues described.

SUGGESTED READING

Morrell MJ. (2003). Reproductive and metabolic disorders in women with epilepsy. *Epilepsia* 44(Suppl. 4), 11–20.

Morrell MJ, Flynn KL, Seale CG et al. (2001). Reproductive dysfunction in women with epilepsy: antiepileptic drug effects on sex-steroid hormones. *CNS Spectrums* 6, 771–786.

Pack AM, Morrell MJ. (2002). Treatment of women with epilepsy. *Semin. Neurol.* 22(3), 289–297.

Pack M, Morrell MJ, Marcus R et al. (2005). Bone mass and turnover in women with epilepsy on antiepileptic drug monotherapy. *Ann. Neurol.* 57, 252–257.

Zupanc ML. (2006). Antiepileptic drugs and hormonal contraceptives in adolescent women with epilepsy. *Neurology* 66(Suppl. 3), S37–45.

34 Unverricht–Lundborg Disease

Danielle M. Andrade, M.D. M.Sc.
and Berge A. Minassian, M.D., C.M., FRCP(C)

CONTENTS

CASE PRESENTATION

The patient is a 20-year-old male referred to the epilepsy clinic for evaluation of seizures and "gait problems." He was the product of an uneventful pregnancy and born to nonconsanguineous parents from an island in the Mediterranean. Developmental milestones were attained at appropriate ages, and he had an otherwise normal development until the onset of seizures. His first generalized tonic–clonic (GTCS) seizure was at the age of 11 years during a hockey game. This type of seizure recurred over the next 7 years, despite treatment with valproic acid (VPA) and clonazepam, with a frequency of 2–6 per year. However, for the last 2 years he has not had any GTCS. There were no clear precipitating factors, except for flashing lights. Since the onset of GTCS, he was also experiencing multifocal, fragmentary, stimulus-sensitive myoclonus. These were sudden, brief, shock-like muscle contractions that at times generalized and interfered with activities of daily living, such as writing, swallowing, speaking, and walking. He also developed ataxia and dysarthria. These symptoms appeared insidiously and progressed slowly. His behavior also changed, and he became very introspective. Over the past few months, he exhibited occasional episodes of aggressiveness, which were never seen before and had not been reported in any of his three healthy male siblings. A clinical evaluation suggested that these episodes were nonepileptic in nature. Finally, the patient dropped out of school after doing poorly for the past two academic years. His neurological exam was significant for the presence appendicular and truncal ataxia, dysarthria, postural and action tremor worse on the right side, and myoclonus. He needed a walker given the severity of the myoclonus. Using a simplified standard myoclonus rating scale, he had a score of 4 (i.e., moderate to severe myoclonus; interference with fine movements, and speech; the patient is able to stand but unable to walk without assistance). A brain magnetic resonance imaging (MRI) and a routine electroencephalogram (EEG) were normal. Genetic testing showed expansion of a dodecamer repeat in the promoter region of the CSTB (or EMP1) gene, thus confirming the diagnosis of Unverricht–Lundborg disease (ULD). The patient was treated with levetiracetam (added to VPA), which lead to clinical improvement (score of 2 using the same myoclonus rating scale) 6 months later.

DIFFERENTIAL DIAGNOSIS

The presence of seizures, myoclonus, and progressive neurological deterioration (ataxia, dysarthria, tremor), in addition to psychiatric symptoms (social withdrawal, depression, anxiety, aggression, and dementia) are strongly suggestive of progressive myoclonus epilepsy (PME). PME represents a group of more than 20 diseases, and although sharing common features, these diseases are distinct in terms of the etiology, pathogenesis, and prognosis. The five most common and better characterized PMEs are Unverricht–Lundborg (ULD), Lafora disease (LD), the neuronal ceroid lipofuscinosis, sialidosis type 1, and myoclonic epilepsy with ragged red fibers (MERRF).

Onset between late childhood and late adolescence is more commonly due to ULD or LD. Both entities transmit via autosomal recessive inheritance. ULD and LD occur worldwide, but are more prevalent in the Mediterranean region. ULD is also seen with increased frequency in Finland, and it may be underdiagnosed in other regions. In both diseases, absence, and complex partial and focal motor seizures may occur in addition to tonic–clonic and myoclonic seizures. However, in LD patients, seizures become increasingly difficult to control with antiepileptic medications. Occipital-onset seizures are suggestive of LD.

Despite progress in treatment, LD is universally fatal within 10–20 years after onset. In contrast, ULD patients may have a normal life span. Notably, in this case, there was a paucity of seizures during the first few years after onset, and the patient has actually been seizure free for the past 2 years. The relatively mild progression 9 years after clinical onset is suggestive of ULD rather than LD.

DIAGNOSTIC APPROACH

History and physical exam: ULD onset is usually between 6 and 15 years. It affects male and females equally. In ULD, generalized tonic–clonic seizures are the first symptom in half the cases, whereas in the other half the opening symptom is stimulus-sensitive myoclonus. Rare ULD patients never develop GTCS. Ataxia, dysarthria, and intentional tremor develop overtime. ULD patients may show a slow decline in intelligence (10 points IQ drop per decade). Their mood is labile, and depression is common.

Blood work: Biochemically, half of ULD patients present with increased urinary excretion of indican, an apparently nonspecific finding that was also reported in other disorders presenting with myoclonus. Reduction of tryptophan and 5-hydroxy-indole-acetic acid was observed in the serum of ULD patients.

Brain MRI is usually normal or it may show nonspecific diffuse atrophy.

Electrophysiology: EEGs early in the course of the disease may be within normal limits in some patients or may show abnormalities even before onset of symptoms in others. Common abnormalities include (1) slow and disorganized background activity; (2) runs of interictal epileptiform activity in the form of fast spike and wave or polyspike and wave discharges, recorded in a generalized or multifocal distribution; and (3) photoparoxysmal response (PPR), which is common and initially presents with a broad range of frequencies. The interictal epileptiform abnormalities diminish with appropriate antiepileptic drug (AED) treatment. Interestingly, epileptiform abnormalities may be seen in patients who never experience clinical seizures. The stimulus-sensitive myoclonus is time locked to the cortical spikes. Cortical hyperexcitability is further demonstrated by giant somatosensory evoked potentials (SSEPs) and an enhanced long-loop (cortical) reflex.

Histopathology: ULD is a neurodegenerative disease without accumulation of storage material or a specific pathologic marker. Previous histopathological studies of patients who died from ULD demonstrated cerebellar granular and Purkinje cell loss, gliosis, and neuronal degeneration of the anterior, lateral, medial, and reticular nuclei of the thalamus. Such studies also reported degenerative changes in the cortex, striatum, mamillary bodies, multiple brainstem nuclei, and ventral gray matter of the spinal cord.

Genetic studies: Prior to the discovery of the gene responsible for the great majority of ULD cases, the diagnosis was based on clinical findings, progression, and on the absence of storage material on peripheral tissue or brain biopsies. The gene responsible for ULD, called *CSTB* or *EPM1*, was identified in 1996. Few patients have point mutations in the coding region of the gene. The most common mutation, by far, is an expansion of a dodecamer repeat in the promoter region of *EPM1*. Normal alleles in this region contain two to three tandem copies of the dodecamer. ULD patients have 30 to 150 copies.

TREATMENT STRATEGY

There is no specific treatment for ULD. Initially, seizures can usually be controlled with VPA. Clonazepam can be used as add-on drug for both seizures and myoclonus. Zonizamide has also shown some benefit in terms of seizure control. High doses of piracetam are used to control myoclonus with moderate efficiency both in the short and long term. Levetiracetam (which is chemically related to piracetam) can also be used to control myoclonus (and possibly tonic–clonic seizures) on a long-term basis, and especially in younger patients. In patients previously treated with piracetam, a switch to levetiracetam was not always possible, and a combination of piracetam and levetiracetam appeared to be the best approach.

Brivaracetam, a new drug chemically related to levetiracetam, is currently being tested in patients with ULD. *N*-acetylcysteine has been shown to improve tremor, gait, and myoclonus. The mechanism of action of *N*-acetylcysteine is poorly understood but is likely related to protection against oxidative stress. Interestingly, *N*-acetylcysteine has been shown to prevent apoptotic death of cultured neuronal cells deprived of nerve growth factor. It is tempting to speculate that *N*-acetylcysteine is protective against the apoptosis recently shown to characterize ULD neurodegeneration. Phenytoin worsens the symptoms in ULD and should always be avoided. It is possible that at least part of the cerebellar neurodegeneration reported in the older literature on ULD was due to phenytoin neurotoxicity. Finally, in one retrospective review, lamotrigine showed lack of efficacy or worsening of myoclonic jerks in five patients with ULD.

LONG-TERM OUTCOME

In the 1960s and early 1970s, the mean survival of ULD patients was 14 years after the onset of symptoms. At present, it is clear that clinical evolution of ULD is greatly influenced by the treatment received. Patients today may have a relatively normal life span. In appropriately managed patients, dementia can be averted, and myoclonus and seizures can be minimized. Avoidance of phenytoin is key to this goal. Phenytoin is exquisitely toxic to ULD patients and is likely a major contributor to the severity of cases described in the past. Still, clinical severity may vary even within families with the same genetic mutation.

A recent study followed ULD patients for over 20 years. The percentage of patients who became wheelchair bound or bedridden varied between 7 and 16%. About 30 to 38% of patients required some form of help for activities of daily living. About 70% were able to walk unassisted, and 30% had a job at last follow-up. In all patients studied, myoclonus was mild at onset, worsened during the first 5 to 10 years after onset, but stabilized subsequently. Remission followed an active phase of epilepsy occurring during the first 10 years of the disease. Serial EEGs suggested that brain activity was influenced by pharmacological treatment. Background activity tended to normalize, and the photoparoxysmal response tended to abate over the years.

PATHOPHYSIOLOGY OF ULD

The gene responsible for ULD was identified using a positional cloning approach. The disease-causing gene, named *CSTB* or *EPM1*, is a 674 base-pair gene that contains 3 exons and localizes to chromosome 21q22.3. *EPM1* encodes a previously known but unmapped cysteine protease inhibitor called cystatin B (CSTB). The identification of two point mutations found in *EPM1* proved that this gene was responsible for ULD. To date, a total of eight different point mutations have been identified in *EPM1*. Some mutations affect conserved splice-site sequences and predict severe splicing defects. Others lead to protein truncation, and yet others affect a conserved amino acid sequence critical for cathepsin (i.e., the target proteases) binding. However, these mutations within the transcriptional unit account for less than 10% of *EPM1* alleles causing ULD. More than 90% of patients have unstable expansion, described previously, of a dodecamer repeat (5′CCCCGCCCCGCG-3′)-175 base-pairs upstream from the translation initiation codon of *EPM1*.

In contrast to some neurodegenerative disorders caused by trinucleotide repeat expansions, there is no correlation between the number of repeats and clinical severity or age of onset in ULD patients. It is suggested that once the dodecamer repeat expands beyond a critical threshold, *EPM1* gene expression is reduced, leading to pathological and physiological consequences.

CSTB functions as an intracellular protease inhibitor able to inhibit cathepsins (lysosomal proteases). In humans, CSTB deficiency leads to neuronal cell loss. However, until recently, it was not clear if the cell loss was due to the *EMP1* mutation or to toxic effects of phenytoin. Recently, it was shown that *epm1* knockout mice never treated with this drug demonstrate apoptotic cell death. These data suggest that CSTB has a role in preventing apoptotic cell death in certain mammalian cells. The mechanisms leading to apoptosis and atrophy observed in humans and mice deficient in CSTB are poorly understood. One proposed mechanism is that cathepsins, which are inhibited by CSTB, directly activate caspases leading to the initiation of apoptosis. Another possible mechanism is that the deficiency in CSTB causes an increase in general proteolysis, thus targeting such unhealthy cells for apoptosis.

How apoptosis could lead to the clinical picture of a hyperexcitable cortex that generates seizures and myoclonus is not clear. It has been suggested that GABA neurons are particularly prone to damage in *cstb*-deficient mice. In these animals, seizure-induced cell death may be responsible for the progressive nature of the disease. It is also possible that the hyperexcitable cortex is caused by an enhancement of tryptophan metabolism in the central nervous system (CNS) along the serotonin and kynurenine pathways.

Finally, it has been shown that, in ULD patients, the thalamostriatal dopaminergic system is dysfunctional. In a small study, there was an improvement of myoclonus in patients receiving a dopamine agonist. This observation may represent a different mechanism responsible for the clinical findings. However, the exact mechanism leading to such deficiency remains to be elucidated.

CLINICAL PEARLS

1. ULD is the most common progressive myoclonus epilepsy.
2. Seizures, ataxia, and stimulus-sensitive myoclonus are the hallmarks of this disease.
3. Seizures in ULD may be easily controlled with antiepileptic medications, and some patients never develop tonic–clonic seizures.
4. Phenytoin is exquisitely toxic to neurons in ULD and should be avoided.
5. With appropriate treatment, most symptoms can be controlled, and life span may be normal.
6. In ULD there is an absence of storage material. Therefore, biopsies are negative (as opposed to the other PMEs with onset between childhood and late adolescence).
7. Expansion of a dodecamer repeat in the promoter region of the *EPM1* gene is responsible for more than 90% of ULD cases.

SUGGESTED READING

Airaksinen EM, Leino E. (1982). Decrease of GABA in the cerebrospinal fluid of patients with progressive myoclonus epilepsy and its correlation with the decrease of 5HIAA and HVA. *Acta Neurol. Scand.* 66: 666–672.

Canafoglia L, Ciano C, Panzica F et al. (2004). Sensorimotor cortex excitability in Unverricht–Lundborg disease and Lafora body disease. *Neurology* 63: 2309–2315.

Franceschetti S, Sancini G, Buzzi A et al. (2007). A pathogenetic hypothesis of Unverricht–Lundborg disease onset and progression. *Neurobiol. Dis.* 25: 675–685.

Koskiniemi M, Donner M, Majuri H, Haltia M, Norio R. (1974). Progressive myoclonus epilepsy. A clinical and histopathological study. *Acta Neurol. Scand.* 50: 307–332.

Koskiniemi M, Toivakka E, Donner M. (1974). Progressive myoclonus epilepsy. Electroencephalographical findings. *Acta Neurol. Scand.* 50: 333–359.

Lalioti MD, Scott HS, Buresi C et al. (1997). Dodecamer repeat expansion in cystatin B gene in progressive myoclonus epilepsy. *Nature* 386: 847–851.

Lalioti MD, Mirotsou M, Buresi C et al. (1997). Identification of mutations in cystatin B, the gene responsible for the Unverricht–Lundborg type of progressive myoclonus epilepsy (EPM1). *Am. J. Hum. Genet.* 60: 342–351.

Magaudda A, Ferlazzo E, Nguyen VH, Genton P. (2006). Unverricht–Lundborg disease, a condition with self-limited progression: long-term follow-up of 20 patients. *Epilepsia* 47: 860–866.

Pennacchio LA, Lehesjoki AE, Stone NE et al. (1996). Mutations in the gene encoding cystatin B in progressive myoclonus epilepsy (EPM1). *Science* 271: 1731–1734.

Pennacchio LA, Bouley DM, Higgins KM, Scott MP, Noebels JL, Myers RM. (1998). Progressive ataxia, myoclonic epilepsy and cerebellar apoptosis in cystatin B-deficient mice. *Nat. Genet.* 20: 251–258.

35 Post-Traumatic Seizures and Epilepsy

Daniel H. Arndt, M.D. and
Christopher C. Giza, M.D.

CONTENTS

CASE PRESENTATION

A previously healthy 12-year-old right-handed female fell while riding horseback and sustained moderate traumatic brain injury (TBI). The primary point of impact was the left temporoparietal area. She suffered a 10-minute loss of consciousness, and her Glasgow Coma Scale (GCS) was 12 (E4, M5, V3) on arrival to the emergency room 2 hours later. Despite her altered state, her neurological examination was nonfocal, and did not show clinical signs of increased intracranial pressure requiring aggressive medical or surgical management. Her head computed tomography (CT) showed two left temporal punctate contusions and a 2 × 3 cm right temporal intraparenchymal hematoma with subarachnoid hemorrhage and subdural hemorrhage along the tentorium. She developed several early posttraumatic seizures (EPTS) that were generalized tonic–clonic convulsions, but did not have status epilepticus. These were controlled with phenytoin. After 2 days, she was discharged home with a postconcussive syndrome, but an otherwise nonfocal neurological examination.

Six months after hospital discharge she began having seizures (late posttraumatic seizures [LPTS], or posttraumatic epilepsy [PTE]) with a different semiology. Although the majority of events were nocturnal, some were diurnal and started with an aura of periorbital pain, which progressed to right arm extension above the head, back arching, and secondary generalization. She experienced a postictal headache. Initially, she had 3–4 seizures over 6 months.

At the onset of her PTE, she was treated with phenytoin and then valproic acid (VPA), antiepileptics with known efficacy for localization-related seizures. However, she continued to have seizures and was transitioned to newer antiepileptics, including lamotrigine, oxcarbazepine, topiramate, and eventually levetiracetam. Unfortunately, these medications did not provide lasting benefit, and she progressed over the next 2 years to daily seizures.

At age 14, she was admitted for an inpatient evaluation. Two seizures were captured on video-EEG (electroencephalogram) as arousals from sleep with right-hand automatisms, then lip smacking, eye blinking, left-head version, left-hand dystonia, and secondary generalization. Electrographically, these episodes showed a broad onset over the right anterior quadrant (F8/T4/T2). She was diagnosed as having complex partial seizures with secondary generalization and a suspected right frontotemporal focus. Magnetic resonance imaging (MRI; see Figure 35.1) showed right temporal encephalomalacia and gliosis consistent with her prior right temporal intraparenchymal bleed. An interictal positron emission tomography (PET) scan (see Figure 35.2) revealed moderate hypometabolism of the right anterior, infero-mesial temporal lobe, and this hypometabolism matched the known MRI findings (see Figure 35.2).

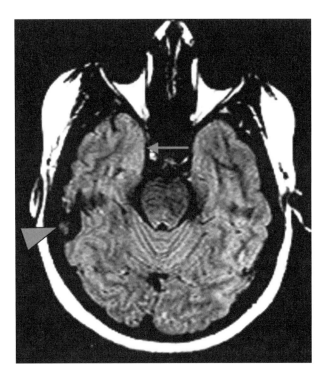

FIGURE 35.1   Fluid-attenuated inversion recovery (FLAIR) sequence MRI showing subtle right anteromedial temporal gliosis (arrow) and right lateral temporal encephalomalacia (arrowhead).

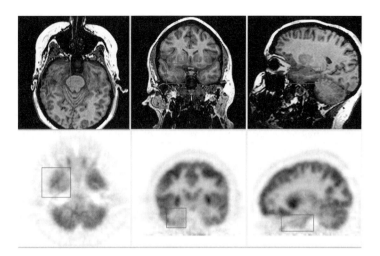

FIGURE 35.2   Fluoro-deoxy-glucose–positron emission tomography (FDG-PET) scan (bottom row) and magnetic resonance imaging–positron emission tomography (MRI-PET) fusion (upper row) demonstrating a matching hypometabolism of the right anterior temporal lobe with the MRI findings shown in Figure 35.1.

Given her medical intractability, and based upon identification of a discrete epileptic focus during her workup, she underwent a focal right temporal surgical resection. Her course was uncomplicated without persistent neurological deficit, and she was maintained on her preoperative antiepileptic regimen. She has remained seizure free on medication for over 6 months since her surgery.

DIFFERENTIAL DIAGNOSIS

The differential for EPTS is broad, including any type of acute symptomatic seizure secondary to intracranial hemorrhage, contusion, edema, electrolyte disturbance, intoxication, or hypoxia/ischemia. An impact seizure is a benign generalized seizure with complete recovery that occurs hyperacutely after trauma, but it remains a diagnosis of exclusion. EPTS that occur during hospitalization may also be (but rarely) due to developing CNS infection or posttraumatic hydrocephalus. Lastly, it is possible that EPTS represent a provoked seizure in a patient with preexisting epilepsy. LPTS have a differential that includes PTE as well as other symptomatic, cryptogenic, and idiopathic localization-related epilepsies.

DIAGNOSTIC APPROACH

The diagnostic approach to posttraumatic seizures differs based on the temporal relationship of the seizure to the inciting TBI. With respect to immediate posttraumatic seizures (IPTS; <24 hours postinjury) or EPTS (<7 days postinjury), a noncontrast head CT scan is essential to rule out hemorrhage, contusion, or other structural lesions. Similar to other initial seizure evaluations, EEG (or even continuous EEG monitoring) may be done acutely in patients with persistent altered mental status (AMS) to rule out subclinical seizures or nonconvulsive status epilepticus (SE). PTS are generally expected to be focal or multifocal in origin (rather than primarily generalized), so initiation of antiepileptic therapy need not be delayed. The incidence of PTS for unselected pediatric TBI ranges from 2.6–9.3%, and the incidence increases significantly with increasing TBI severity. Interestingly, EEG during the acute post-TBI time period has *not* been shown to reliably predict the long-term development of PTE. MRI is also not essential acutely; however, it can be helpful in identifying lesions not visible on CT scan, including diffuse axonal injury. In patients with persistent AMS, MR spectroscopy can add additional, independent prognostic value.

In patients with LPTS/PTE (recurrent, unprovoked seizures > 7 days post-TBI), routine EEG and MRI are indicated. PTE represents 25% of symptomatic epilepsy in the general population. EEG may again be used acutely to rule out subclinical seizures or nonconvulsive SE, and interictal epileptiform discharges after the acute period may more reliably reflect the development of PTE. Video-EEG may be useful in distinguishing epileptic from nonepileptic paroxysmal events in low-functioning post-TBI patients. Routine MRI may demonstrate chronic, focal changes such as hemosiderin deposition or gliosis.

TREATMENT STRATEGY

Treatment for PTS in general is aimed at preventing partial onset seizures, given that the seizures originate from focal or multifocal injuries to the brain. Randomized controlled treatment trials for PTS in children are few, and treatment recommendations differ for EPTS and LPTS/PTE.

Most patients with moderate–severe TBI receive phenytoin for 1 week aimed at preventing EPTS, although this has not been explicitly studied in a pediatric population. The largest, randomized, controlled treatment trial for PTS found that the incidence of EPTS in teenagers and adults with moderate–severe TBI was 3.6% in patients receiving phenytoin prophylaxis versus 14.2% of those assigned to placebo. However, by one year after TBI, there was no significant benefit of prophylaxis, and side effects of treatment became problematic. Alternatives for EPTS prophylaxis include other antiepileptic drugs (AEDs) with effectiveness against localization-related epilepsies. Phenobarbital can be used for acute prophylaxis, but its sedating effects may mask changes in mental status, making it less than ideal after TBI. VPA may also be effective, but has a higher risk of adverse effects, including coagulopathy, which is problematic in TBI cases with intracranial hemorrhage. A recent clinical literature review also supported carbamazepine as a treatment alternative for EPTS prophylaxis after TBI. Preclinical studies suggest that both levetiracetam and topiramate are less neurotoxic to the developing brain, with the added potential of providing antiepileptogenic effects. Oxcarbazepine, lamotrigine, and zonisamide are other second-generation alternatives, but, similarly to topiramate, are not available intravenously. Posttraumatic status epilepticus (PTSE) occurs in 5–10% of children, more commonly with EPTS than LPTS, and is treated with parenteral benzodiazepine administration followed by parenteral prophylactic medications.

As indicated, there is no clear demonstration that anticonvulsant prophylaxis beyond 1 week post-TBI provides any meaningful reduction in the risk of LPTS/PTE. Should it occur, LPTS/PTE can be managed using antiepileptics that are typically effective for localization-related epilepsies. It is not clear whether PTE is more difficult to control than other types of localization-related epilepsy, but it is certainly well known that PTE can be intractable to medications.

LONG-TERM OUTCOME

EPTS have been associated with poorer outcomes, higher mortality rates, and increased risk of neurologic sequelae. Poorly controlled seizures are reported to be the second most common, avoidable neurologic factor contributing to death after TBI. The first being hypoxia/hypotension.

Pediatric EPTS are a strong risk factor for the development of LPTS/PTE. The expected incidence of LPTS/PTE is 11.6–41.3% (3–9-fold increased risk) in patients with EPTS. However, a large population-based study did not confirm this correlation in children, although it was seen in adults. Any proposed relationship between EPTS and LPTS is also tempered by the fact that studies showing the efficacy of acute seizure prophylaxis failed to show any significant reduction in the rate of LPTS/PTE in the treated patients.

Only 2 pediatric studies incorporated an objective outcome scale, the Glasgow Outcome Scale (GOS), to assess the effect of EPTS on the long-term outcome of TBI patients. Both suggested that poorer outcomes were more likely in those with EPTS versus those without.

Late PTS/PTE outcome is less favorable in that it carries the same risks as with other chronic symptomatic epilepsies. Furthermore, post-TBI neurobehavioral impairments can put these patients at higher risk for seizure recurrence due to noncompliance, and for repetitive TBI due to aggression, risk-taking, and impulsivity. Poorly controlled seizures themselves increase the risk for future TBI and other injuries.

PATHOPHYSIOLOGY

The pathophysiology underlying PTS changes over time, as indicated in Figure 35.3. EPTS result from primary or secondary brain damage and may indicate the presence of intracranial hemorrhage, worsening cerebral edema, hypoxia, concomitant intoxication, or other ominous factors associated with TBI. Any PTS may itself induce secondary brain injury by increasing metabolic requirements, elevating intracranial pressure (ICP), inducing cerebral hypoxia, exacerbating indiscriminate neurotransmitter release, dropping blood pressure/cerebral perfusion, and/or elevating temperature.

LPTS have been attributed to different neurobiological processes, including delayed cell death, excitatory–inhibitory imbalance, toxic effects of hemosiderin deposition, and chronic astrogliosis/scarring. Although no antiepileptogenic agents have yet been proven clinically effective, these processes serve as potential future targets for therapies designed to prevent the development of LPTS/PTE.

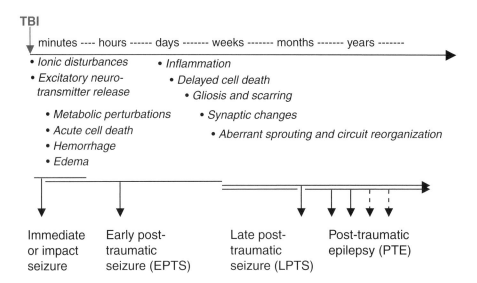

FIGURE 35.3 The temporal relationship between post-TBI pathophysiology and posttraumatic seizure subtypes.

CLINICAL PEARLS

1. Posttraumatic seizures are classified as early (EPTS; < 7 days of TBI) or late (LPTS; >7 days of TBI), whereas posttraumatic epilepsy (PTE) is characterized by recurrent (i.e., >1), unprovoked LPTS.
2. Strong risk factors include severe TBI, acute CT scan abnormalities, and focal neurologic signs. Intermediate risk factors include moderate TBI, depressed skull fractures, prolonged loss of consciousness, and/or posttraumatic amnesia.
3. The workup for EPTS includes CT scan acutely, whereas EEG and MRI may be done for further evaluation.
4. Patients with moderate-to-severe TBI may receive phenytoin prophylaxis against EPTS for 7 days. Longer-term prophylaxis has not been shown to be effective.
5. If PTE develops, these patients may be treated similarly to those with other localization-related epilepsies.

SUGGESTED READING

Annegers J, Hauser W, Coan S, Rocca, W. (1998) A population-based study of seizures after traumatic brain injuries. *N. Engl. J. Med.* 338: 20–24.

Beghi E. (2003) Overview of studies to prevent posttraumatic epilepsy. *Epilepsia* 44(s10): 21–26.

Bittigau P, Sifringer M, Genz K, Reith E et al. (2002) Antiepileptic drugs and apoptotic neurodegeneration in the developing brain. *Proc. Natl. Acad. Sci. USA* 99(23): 15089–15094.

Elvidge, A. (1939) Remarks on post-traumatic convulsive state. *Trans. Am. Neurol. Assoc.* 65: 125–129.

Hudak A, Trivedi K, Harper C et al. (2004) Evaluation of seizure-like episodes in survivors of moderate and severe traumatic brain injury. *J. Head Trauma Rehabil.* 19(4): 290–295.

Statler K. (2006) Pediatric Posttraumatic Seizures: Epidemiology, putative mechanisms of epileptogenesis and promising investigational progress. *Dev. Neurosci.* 28: 354–363.

Temkin N, Dikmen S, Wilensky A, Keihm J, Chabal S, Winn H. (1990) A randomized, double-blind study of phenytoin for the prevention of post-traumatic seizures. *N. Engl. J. Med.* 323: 497–502.

36 Reflex Epilepsy

Michael C. Kruer, M.D. and
Colin M. Roberts, M.D.

CONTENTS

CASE PRESENTATION

An 8-year-old developmentally normal boy is brought to the clinic by his mother for episodes of staring and unresponsiveness. She reports that, as early as age 4 years, he would stare out the window of the car at the sun and blink repeatedly. When questioned about this behavior, he reported that "it felt good." One particularly sunny day, while riding in the car, he continued to stare at the sun until he developed myoclonic jerking that ultimately progressed to whole-body tonic stiffening lasting several minutes. He was taken to a local emergency room where an EEG with photic stimulation evoked a second seizure. Shortly thereafter, he began to engage in an unusual stereotyped behavior. Throughout the day, he would look upwards at the sun or towards a light, and wave a hand in front of his face with fingers spread. As he did this, he would blink several times and then become unresponsive for a few seconds. After a large shiver, he would return immediately to baseline and resume his normal activity.

This behavior had increased in frequency over several months. When most severe, it would occur every 5–10 minutes, perhaps more often if playing in direct sunlight. He will not stop if redirected, but can suppress the action in order to complete other tasks. If one hand is restrained, he will use the other with the same effect. He is aware of the episodes, and does not find them disturbing, but rather reports a pleasurable sensation associated with them. The patient has had no other convulsions, and no episodes of staring or myoclonus have been observed independent of his hand waving. His EEG shows a normal background pattern. When he waves his hand before his eyes, however, an irregular generalized spike and slow-wave discharge follows (Figure 36.1). Clinically, during this time he exhibits facial myoclonus and staring. No independent spontaneous epileptiform features are noted. Photic stimulation elicits a generalized photoconvulsive response at flash frequencies ranging from 12 to 20 Hz. When a blue filter is placed over the strobe, this response is extinguished.

DIFFERENTIAL DIAGNOSIS

Diagnostic considerations include nonreflex epilepsies that the child's parents or caregivers have erroneously associated with certain activities, particularly if only a few seizures have occurred. Proprioception or praxis-induced seizures may be confused with kinesigenic dyskinesias or with self-stimulatory behaviors in children with pervasive developmental disorders. Language and/or reading epilepsy may be confused with dyslexia, tics, or stuttering, and, in fact, acquired stuttering may be the only manifestation of expressive language-reflex epilepsy. Syncope or breath-holding spells may be considered in some cases. Finally, paroxysmal nonepileptic seizures are a consideration, particularly in adolescents.

Reflex seizures are seizures in which an afferent stimulus produces an ictal response. There are several different types of reflex seizures, and in general, they can be designated as "simple" or "complex," based on the nature of the provoking stimulus. Simple reflex seizures tend to involve relatively uncomplicated stimuli, such as the visual experience of a distinct geometric pattern, whereas complex reflex seizures tend to require an exacting combination of sensory stimuli and cognitive processing, typically involving association cortex.

Examples of simple reflex epilepsies include the flicker or flash-induced seizures described earlier, as well as seizures induced by bathing ("hot water epilepsy"), whereas complex reflex seizures include thinking-induced ("noogenic") seizures and musicogenic epilepsy, among numerous others. Note that the designations "simple" or "complex" do not refer to the type of seizure that the stimulus produces but rather the complexity of the inducing stimulus. Both simple and complex reflex seizures are thought to occur by stimulation of an area of the cerebral cortex within or functionally connected to an epileptogenic zone.

Both focal and generalized reflex seizures have been described, although not typically within the same syndrome. Each type of reflex epilepsy has a typical constellation of associated seizure semiologies. Reflex seizures may occur in isolation or concurrently with other, unprovoked seizure types. Although some reflex seizure variants occur only rarely, others are quite common. For example, the morning movement-induced myoclonic seizures of juvenile myoclonic epilepsy (JME) appear to represent a form of praxis-induced seizures. Similarly, the photoconvulsive response (PCR) seen in many idiopathic generalized epilepsies with intermittent photic stimulation exists along the continuum of reflex seizures. The numerous types of reflex epilepsy are briefly summarized in Table 36.1.

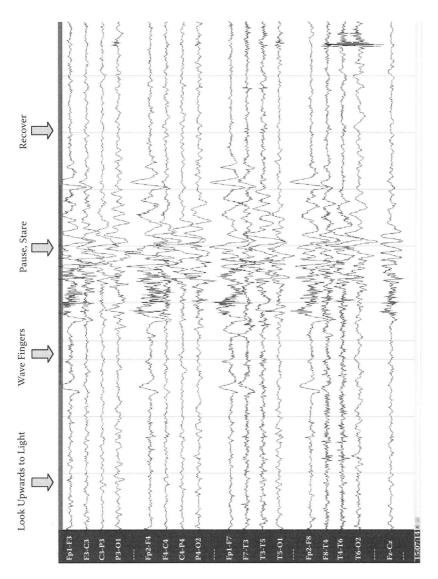

FIGURE 36.1 An 8-year-old self inducing a photoconvulsive seizure by waving his fingers before his eyes. Note how the physical induction precedes the polyspike discharge and clinical absence.

TABLE 36.1
The reflex epilepsies

Epilepsy types	Epilepsy	Origin	Seizure types	Notes
Somatosensory Touch Proprioception	Partial	Somatosensory cortex	Tonic Focal motor	Often have Jacksonian component "Hot water," "eating," "toothbrushing" variants Walking-induced variant
Memory	Partial	Limbic	Focal motor Complex-partial	
Audiogenic Musicogenic Speech-induced	Partial	Primary auditory cortex Temporal, limbic cortex	Focal motor Complex-partial	LGI1-associated; prominent in Fragile X syndrome
Photosensitive Flash Geometric pattern Color	Generalized	Primary visual cortex	Myoclonic Absence Tonic–clonic	Associated with stimulus induction
Thinking	Generalized	Nondominant parietal	Myoclonic Absence Tonic–clonic	"Arithmetic," "mental imagery," "chess," "Rubik's cube" and "Mah Jong" variants, among numerous others
Praxis	Generalized	Nondominant parietal, prefrontal	Myoclonic Absence Tonic–clonic	Utensil-induced variant
Reading	Generalized	Temporoparietal	Myoclonic Absence Tonic–clonic	Likelihood of seizure may increase with duration or complexity of material
Startle	Generalized	Supplementary somatosensory, prefrontal	Tonic Myoclonic	Seen typically in children with mental retardation and/or diffuse cortical anomalies

The boy presented in the vignette has photic flicker-induced seizures, and his presentation is consistent with simple reflex epilepsy. He self-induces his seizures via the stroboscopic effect of light seen through his waving fingers. These seizures can be a challenge to treat, as standard antiepileptic treatment may not fully extinguish the photoconvulsant effect. In addition, the wide range of environmental light triggers, combined with the patient's pleasurable secondary gain, can make for a uniquely challenging scenario.

Flicker-induced seizures may be seen in generalized epilepsy syndromes, as well as in focal occipital lobe epilepsies. In addition, a number of varieties of photic stimulation-sensitive seizures have been described, including color (wavelength) and frequency-sensitive seizures, seizures induced by certain geometric patterns, seizures triggered by stepping into the light after emerging from a dark area, and seizures triggered by diverting one's attention away from an item of visual interest ("fixation-off" seizures). An overlap of photosensitive epilepsy with migraine may be seen, and an association of headache with occipital spiking has been described within this association.

Self-induced reflex seizures were first described by Radovici (1932). There is a peculiar tendency of some patients with photosensitive epilepsy to actively seek out the very light that, in turn, may provoke an attack. The motivation of patients with this so-called "sunflower syndrome" may perhaps be understood with the realization that photosensitive seizures may be pleasurable for some patients, and patients will thus engage in hand-waving, finger-flapping, or staring at the television in order to purposefully induce seizures. Many children that stimulate their own photosensitive seizures may have comorbid mental retardation, but a significant proportion of them are cognitively normal. Intentional provocation of seizures in order to avoid school or a stressful life event has been reported. Some patients also report that they may induce seizures at a convenient time in order to take advantage of the relative refractory period that follows.

DIAGNOSTIC APPROACH

A careful history is necessary when reflex epilepsy is suspected in order to determine how reliably the stimulus provokes the seizure, and whether other subtle seizure types may be present. In difficult cases, video-EEG (electroencephalogram) telemetry, along with presentation of the offending stimulus, may be invaluable. Neuroimaging may be useful, especially in those reflex epilepsies that have a focal component based on history or EEG findings.

TREATMENT AND LONG-TERM OUTCOME

Treatment of the reflex epilepsies may involve avoidance of the offending stimulus and/or use of antiepileptic drugs. Avoidance alone may be reasonable and sufficient to control seizures, but may be impossible in some situations (i.e., noogenic epilepsy or reading epilepsy). In cases where avoidance alone is insufficient, antiepileptic drugs (AEDs) may be employed. Given the prominent myoclonic nature of many reflex seizures, valproic acid (VPA) is most commonly used, and there is a

CLINICAL PEARLS

1. Reflex seizures can be induced by a variety of stimuli, and are classified based on the type and complexity of the stimulus.
2. Avoidance of triggering stimuli may allow control of seizures. Antiepileptic medications should be used for those patients with continued seizures.
3. For patients with lesional reflex epilepsy, epilepsy surgery may make intractable patients seizure free.

SELECTED READING

Binnie CD. (1988) Self-induction of seizures: the ultimate non-compliance. *Epilepsy Res.* Suppl. 1: 153–158.

Bruhn K, Kronisch S, Waltz S, Stephani U. (2007) Screen sensitivity in photosensitive children and adolescents: patient-dependant and stimulus-dependant factors. *Epileptic Disord.* 9(1): 57–64.

de Haan GJ, Trenité DK, Stroink H, Parra J, Voskuyl R, van Kempen M, Lindhout D, Bertram E. (2005) Monozygous twin brothers discordant for photosensitive epilepsy: first report of possible visual priming in humans. *Epilepsia* 46(9): 1545–1549.

Radovici A, Misirliou V, Gluckman M. (1932) Epilepsie reflexe provoquee par excitations des rayons solaires. *Revue Neurologique* 1: 1305–1308.

Siniatchkin M, Groppa S, Jerosch B, Muhle H, Kurth C, Shepherd AJ, Siebner H, Stephani U. (2007) Spreading photoparoxysmal EEG response is associated with an abnormal cortical excitability pattern. *Brain* 130(pt. 1): 78–87.

Szabó CA, Narayana S, Kochunov PV, Franklin C, Knape K, Davis MD, Fox PT, Leland MM, Williams JT. (2007) PET imaging in the photosensitive baboon: case-controlled study. *Epilepsia* 48(2): 245–253.

Valenti MP, Rudolf G, Carré S, Vrielynck P, Thibault A, Szepetowski P, Hirsch E. (2006) Language-induced epilepsy, acquired stuttering, and idiopathic generalized epilepsy: phenotypic study of one family. *Epilepsia* 47(4): 766–772.

37 Autosomal Dominant Nocturnal Frontal Lobe Epilepsy

Kevin Chapman, M.D.

CONTENTS

CASE PRESENTATION

A 10-year-old right-handed male without a significant past medical history was referred for evaluation of possible nocturnal seizures. His seizures began at 7 years of age and have occurred almost exclusively during sleep, usually within 2 hours of falling asleep. The stereotyped episode involves rocking and bicycling motions, gagging, rolling of the eyes, and profuse sweating lasting approximately 2 minutes. These episodes have occurred on average three times per week, but recently have become more frequent. He has had two similar episodes that have occurred while awake. The patient's previous evaluation included two routine sleep-deprived EEGs and a 1.5T noncontrast MRI of the brain, all of which were interpreted as normal. Other laboratory tests included a complete blood count and comprehensive metabolic profile, which were also normal. His physical and neurologic examination is unremarkable, but his family history is significant for a maternal aunt and brother with epileptic seizures since childhood that were primarily nocturnal.

Ultimately, the patient was admitted for continuous video-EEG (electroencephalogram) monitoring to characterize the episodes of concern. During this study, the patient had three typical clinical events identified by the family. Clinically, the patient would arise, sit up in bed, rock, and look around the room, but not respond. He made some occasional nonsensical vocalizations. The events lasted from 47 to 83 seconds. During the events, the EEG demonstrated a change from a normal stage II sleep recording to a generalized frontal dominant rhythmic 3 Hz high-amplitude slow activity (See Figure 37.1). This activity lasted half a minute and gradually waned to a normal background rhythm without any focal slowing. Based on this video-EEG study, the diagnosis of nocturnal frontal lobe epilepsy was made, and the patient was started on carbamazepine. Genetic testing demonstrated a mutation in the *CHRNA4* gene, consistent with the diagnosis of autosomal dominant nocturnal frontal lobe epilepsy (ADNFLE). The patient responded well to carbamazepine, and would experience only rare seizures after missed doses of medication.

DIFFERENTIAL DIAGNOSIS

The unusual behaviors seen in patients with ADNFLE can make their diagnosis challenging. Depending on the site of origin, patients may exhibit complex automatisms, such as bicycling or rocking, vocalizations, dystonic posturing, or clonic activity, but may retain awareness. The differential diagnosis of paroxysmal nocturnal events involves distinguishing parasomnias from true epileptic seizures. Pediatric parasomnias (e.g., night terrors, sleep walking, and confusional arousal) occur in up to 6% of the population, and may be difficult to differentiate from ADNFLE. Night terrors (aka, pavor nocturnus) often occur during slow-wave sleep in the first third of the night. The patients will often make a loud cry and appear frightened. They may have thrashing movements, as if defending themselves, but these are typically not stereotyped as in ADNFLE. During the events, they are often unarousable, but after the events may recall a frightening dream.

Confusional arousals often begin with simple movements or moaning that gradually progress to a more agitated and confused state that may last 5–15 minutes. The patients are often difficult to arouse and have little recollection of the event. Sleepwalking may occur in children, but is typically associated with a calm demeanor that differs from the parasomnias or seizures described earlier. REM sleep behavior disorder may also have similar presentations, but occurs later in the night and is uncommon in children. Polysomnograms demonstrate a lack of atonia during REM sleep in these patients.

Useful clues to help differentiate ADNFLE from parasomnias include the presence of stereotyped behaviors, events occurring during periods of wakefulness, a history of clear seizures, a family history of epilepsy, later age of onset (parasomnias usually begin between 4 and 6 years of age), and multiple events per night. Seizures are less likely to occur during REM sleep, and more often arise during transition from sleep to waking. The EEGs in ADNFLE often lack clear abnormalities, and may be normal despite multiple seizures per night.

DIAGNOSTIC APPROACH

It is often necessary to have a high index of suspicion for ADNFLE, as patients afflicted with this disorder are often misdiagnosed for many years as parasomnias. ADNFLE is an idiopathic partial epilepsy with a mean age of onset of 11 years and an equal male-to-female ratio. The medical history is often of utmost importance in proper diagnosis. The events are typically brief (<30 seconds) and may occur multiple times per night, something which is unusual for parasomnias. Secondarily generalized tonic–clonic seizures may also occur in this syndrome. ADNFLE is transmitted in an autosomal dominant mode, with a penetrance of about 70%. There is marked variability within affected families, with some members being more severely affected than others. Individuals appear to have normal intellect, but careful studies of neuropsychological functioning are lacking. The video-EEG in ADNFLE can often be unrevealing and may cast doubt in the diagnosis, but remains an effective tool for differentiating seizures from parasomnias. Patients may have interictal abnormalities seen primarily in sleep that are frontal or bifrontal, such as spikes or focal slowing. However, more than half of ADNFLE patients do not exhibit interictal abnormalities. Typical ictal electrographic changes consist of diffuse attenuation or rhythmic slowing over the anterior regions (See Figure 37.1). Some patients may lack any clear evolving ictal change. The seizures primarily occur out of non-REM sleep, but may occur while awake in 35% of patients. About 10% of patients lack seizures during wakefulness and have normal EEGs. Polysomnographic recording may help differentiate the epileptic events from nonepileptic parasomnias, such as REM behavior disorder. Genetic testing can be invaluable in confirming the diagnosis of ADNFLE. Mutation analysis of the *CHRNA4* and *CHRNB2* genes is commercially available. However, other genes are likely responsible for the majority of ADNFLE cases as these two genes account for only about 10–20% of cases.

TREATMENT STRATEGY

ADNFLE often responds well to antiepileptic drug therapy. Carbamazepine appears to be particularly effective in nearly two-thirds of patients. Other medications, such as phenytoin, clonazepam, valproic acid, lamotrigine, and acetazolamide, have been used with varying success. Interestingly, improved seizure control has been reported with tobacco use, presumably through the nicotine intake associated with smoking. The utility of transdermal nicotine in the treatment of refractory ADNFLE has been studied in one patient in a double-blind manner with significant improvement.

LONG-TERM OUTCOME

Although many patients respond to antiepileptic drug therapy, one-third of patients may continue to be refractory to various medical treatments. Long-term outcome data on patients with ADNFLE are lacking, but it has been observed that many patients experience seizures well into adulthood.

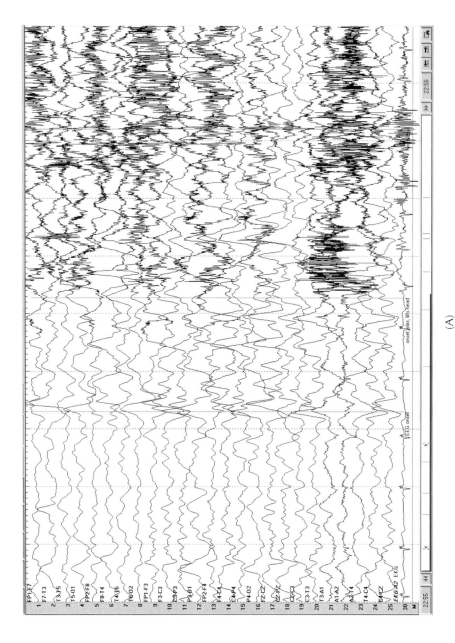

(A)

FIGURE 37.1 A, B   Electroencephalogram demonstrating a nocturnal seizure with rhythmic delta activity evolving over the frontal head regions.

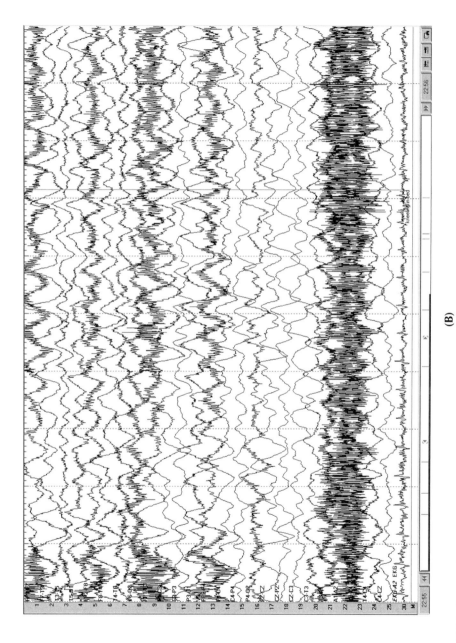

(B)

FIGURE 37.1 A, B   (continued)

PATHOPHYSIOLOGY/NEUROBIOLOGY OF DISEASE

Autosomal dominant nocturnal frontal lobe epilepsy was the first idiopathic epilepsy syndrome that was linked to a specific genetic mutation. Its discovery opened the door for research into the molecular genetics of epilepsy, and further validated a class of disorders known as "ion channelopathies." ADNFLE is due to a mutation in a gene encoding a subunit of the neuronal nicotinic acetylcholine receptor (nAChR). Thus far, three separate loci have been identified: 20q13 (ENFL1), 15q24 (ENFL2), and 1q21 (ENFL3). Four separate mutations in the *CHRNA4* gene that encodes the α4 subunit of the nAChR have been discovered, and correspond to the ENFL1 locus. ENFL3 has been mapped to the *CHRNB2* gene encoding the β2 subunit, whereas the gene associated with *ENFL2* has yet to be identified. The mutations in these two genes account for less than 15% of the patients diagnosed with ADNFLE, suggesting other causative genes.

Cholinergic projections from the basal forebrain nuclei are found throughout the cortex and hippocampus, and likely play a role in learning and memory. The cholinergic projections from the laterodorsal and pedunculopontine tegmentum, which are part of the reticular activating system, play an important role in the regulation of sleep and arousal through the thalamocortical system. Depolarization of these nuclei causes cortical activation and desynchronization of the EEG necessary for wakefulness. Nicotine stimulates these projections and can enhance cortical activation.

The neuronal nAChR is a pentameric ligand-gated ion channel, composed of various combinations of alpha and beta subunits, which can be opened by nicotine or acetylcholine. Permeability to various cations is dictated by the subunit combination, with the most common subtype in the mammalian brain consisting of α4β2 subunits. nAChRs are found primarily in the presynaptic nerve terminals. Their activation leads to a brief depolarizing excitatory potential, and they can allow calcium influx leading to increased neurotransmitter release.

The pathogenic mechanisms through which mutations in nAChR subunits cause nocturnal frontal-lobe seizures remain unclear. Electrophysiological studies of the specific nAChR mutations have demonstrated varying alterations of channel function, but increased sensitivity to acetylcholine seems common to all mutations. This increased sensitivity may allow enhanced cortical GABAergic inhibition that may cause inhibitory hypersynchronization and seizures. It remains unclear why characteristically focal-onset (i.e., frontal-lobe) seizures occur when the mutation is widely expressed throughout the brain. The important role that nAChRs play in regulating sleep may provide clues to understanding the pathophysiology of nocturnal seizures.

CLINICAL PEARLS

1. ADNFLE is characterized by seizures with unusual semiology, and misdiagnoses with parasomnias are common.
2. Most patients respond well to carbamazepine monotherapy, but 30% may remain intractable to medical therapy.

3. Video-EEG evaluations may demonstrate interictal and ictal patterns in the frontal regions, but some studies may be normal.
4. Genetic testing for ADNFLE is commercially available, but not all causative mutations have been discovered.
5. The underlying pathophysiology of ADNFLE remains unclear, and is an active area of research.

SUGGESTED READING

Brodtkorb E, Picard F. (2006). Tobacco habits modulate autosomal dominant nocturnal frontal lobe epilepsy. *Epilepsy Behav.* 9(3), 515–520.

Combi R, Dalpra L, Tenchini ML, Ferini-Strambi L. (2004). Autosomal dominant nocturnal frontal lobe epilepsy: a critical overview. *J. Neurol.* 251, 923–934.

Dani JA, Bertrand D. (2007). Nicotinic acetylcholine receptors and nicotinic cholinergic mechanisms of the central nervous system. *Annu. Rev. Pharmacol. Toxicol.* 47, 699–729.

Klaassen A, Glykys J, Maguire J, Labarca C, Mody I, Boulter J. (2006). Seizures and enhanced cortical GABAergic inhibition in two mouse models of human autosomal dominant nocturnal frontal lobe epilepsy. *PNAS* 103(50), 19152–19157.

Mason TB, Pack AI. (2007). Pediatric parasomnias, *Sleep* 30(2), 141–151.

Oldani A, Zucconi M, Asselta R. et al. (1998). Autosomal dominant nocturnal frontal lobe epilepsy: a video-polysmonographic and genetic appraisal of 40 patients and delineation of the epileptic syndrome. *Brain* 121, 205–223.

Scheffer IE, Bhatia KP, Lopes-Cendes I. et al. (1994). Autosomal dominant frontal epilepsy misdiagnosed as sleep disorder. *Lancet* 343, 515–517.

Scheffer IE. (2000). Autosomal dominant nocturnal frontal lobe. *Epilepsia* 41(8), 1059–1060.

Willoughby JO, Pope KJ, Eaton V. (2003). Nicotine as an antiepileptic agent in ADNFLE. *Epilepsia* 44(9), 1238–1240.

Index

A

AAN. *See* American Academy of Neurology
Abruptio placentae, 75
Acetazolamide, 249, 275
Acid-glycoprotein concentrations, 16
Acidosis, potential side effect of KD, 32
ACTH. *See* Adrenocorticotropic hormone
Acute infection, 199
Acyclovir, 137–138
ADHD. *See* Attention deficit hyperactivity
    disorder
Adolescents, 237–279. *See also* Child; Infants;
    Neonates
    autosomal dominant nocturnal frontal lobe
        epilepsy, 273–279
    females, 245–250
    juvenile myoclonic epilepsy, 239–243
    post-traumatic seizures, epilepsy, 257–263
    reflex epilepsy, 265–272
    Unverricht-Lundborg disease, 251–256
Adrenocorticotropic hormone, 4, 111–112, 114,
    222, 228
Adult nervous system, developing nervous
    system, seizure threshold,
    compared, 9
Afebrile seizures, 89–90
Affective disorders, with epilepsy, 11
Age-specific maintenance dosing, 23
Aggression, 252
Albumin, 16
Alcohol use, 5, 240, 246, 248
Alkalinization, potential side effect of KD, 31
Altering quality of light source, 270
Alternative ketogenic diet, 32–34
American Academy of Neurology, 59, 142
Amniotic fluid, meconium passage into, 75
Amphetamine use, 240
Anatomical resection, 54
Angioma
    leptomeningeal, 155–156
    venous, without cerebral angiomatosis, 155
Angiomyolipomas, renal, 128
Anovulatory cycles, 247
Antiepileptic medications, 15–27
    acid-glycoprotein concentrations, 16
    age-related variables, 15
    age-specific maintenance dosing, 23
    albumin, 16

aqueous, lipid solubility, 15
carbamazepine, 17–19, 23
clobazam, 17, 20
clonazepam, 17, 20
cytochrome P450, 16
diazepam, 17–19
elimination by cytochrome P450-dependent
    metabolism, 18–19
elimination by hepatic metabolism, renal
    excretion, 16–18, 21–22
elimination by mixed cytochrome P450,
    uridine glucuronosyltransferase, other
    metabolic pathways, 20–21
elimination by uridine
    glucuronosyltransferase-dependent
    hepatic metabolism, 19–20
ethosuximide, 17, 21, 23
felbamate, 17, 21, 23
gabapentin, 17–18, 23
glomerular filtration rates, 16
hepatic biotransformation to metabolites, 16
influence of age on hepatic metabolism, 16
ionization constant, 15
lamotrigine, 17, 19–20, 23
levetiracetam, 17, 21–23
lorazepam, 17, 19–20
metabolic elimination, 17
molecular size, 15
oxcarbazepine, 17, 22–23
pharmacokinetic properties, 17
phenobarbital, 15, 17, 22–23
phenytoin, 17–19, 23
pregabalin, 17–18, 23
protein binding, 16
renal blood flow, 16
synaptogenesis, interactions, 10
through reproductive years, 246–248
tiagabine, 17, 21, 23
topiramate, 17, 22–23
tubular secretion, 16
uridine diphosphate, 16
uridine glucuronosyltransferase, 16
valproic acid, 16–17, 21, 23
vigabatrin, 17–18, 23
volume of distribution, 15
zonisamide, 17, 22–23
Antipyretics, 107
Anxiety, 249, 252
Aplastic anemia, 194